resistant materials technology

revised edition

Ted Cosway ■ Melanie Fasciato ■ Hilary Felstead
David Macklin ■ Tristram Shepard

First published in 1997 by
Stanley Thornes (Publishers) Ltd

This edition published in 2001 by:
Nelson Thornes Ltd
Delta Place
27 Bath Road
CHELTENHAM,
GL53 7TH
United Kingdom

A catalogue record for this book is available from the British Library.

ISBN 0 7487 6083 0

01 02 03 04 / 10 9 8 7 6 5 4 3 2 1

Designed and typeset by Carla Turchini
Picture research by johnbailey@axonimages.com and Jennie Karrach
Artwork by Andrew Loft, Hardlines and Tristram Ariss
Printed and bound in Italy by Canale

The authors would like to acknowledge the influence of the following books and magazines and recommend them as sources of further information for teachers and pupils:

Design and Make It! Graphic Products, Stanley Thornes (Publishers) Ltd (1996) – revised edition, Nelson Thornes 2001
GCSE Craft, Design and Technology, R. Kimbell, Thames/Hutchinson (1987)
Green Design, Dorothy Mackenzie, Laurence King (1991)
Mastering Manufacturing, Gordon Mair, Macmillan (1993)
New Designer, (Termly magazine for GCSE Design and Technology), Philip Allen Publishers (Tel: 01869 338652)

Thanks are due to Jet Mayor, Russ Jones and Roy Ballam for their contributions to the revised edition.

Sources of Further Information

British Standards Institute, Maylands Avenue, Hemel Hempstead HP2 4SQ (Tel: 01442 230442)
The Crafts Council, 44a Pentonville Road, Islington, London N1 9BY (Tel: 020 7278 7700)
Intermediate Technology, Myson House, Railway Terrace, Rugby CV21 3HT

The publishers are grateful to the following for permission to reproduce photographs or other illustrative material:

Adams Picture Library: pp. 69 (top), 126 (bottom right)
Addis: p. 133 (top)
AKG Photo London: pp. 33 (bottom inset – Kurt Hartman), 46 (bottom right), 47 (centre – Erich Lessing)
Ancient Art and Architecture Collection: p6 (top and bottom left)
Andra Nelki: p. 56, 79 (top and bottom)
Angela Lumley: p. 76 (top)
Anne Finlay: pp. 30 (top), 61 (top)
Bentley pp. 39, 96–97, 118, cover
BICC: pp. 132–133 (top centre)
Bridgeman Art Library/Private Collection: p. 6 (bottom right – Stapleton Collection, UK), 7 (bottom left – Bonhams, London, middle right – Private Collection), 46 (bottom left)
Bruce Coleman Ltd: p. 73 (centre – Hans Reinhard)
Cannon Shelley: p. 73 (top left)
Casio: p. 30 (bottom)
Centre for Rapid Prototyping in Manufacturing, University of Nottingham: p. 119 (centre right and bottom right and centre)
Channel 4 Learning: pp. 130 (bottom), 135 (bottom left, centre right and centre inset), 138
Christies, Bristol: pp. 134, 135 (top right), cover
Christies' Images, London: p 7 (middle left – Wiessenhof MR 20 armchair by Mies Van der Rohe, Bamberg Metallwerkstatten, bottom right – Peter Max), 49 (bottom)
Cookson Precious Metals Birmingham: p. 62
Crafts Council: pp. 55 (bottom left – Peter Chatwin/Pamela Martin), (bottom right – K. Whiterod), 151 (bottom right – M. Stevenson)
Dyson: p 8 (bottom right)
De Vere Hotel, Swindon: pp. 28–29 (centre)
et archive: pp. 32 (bottom left and right), 46 (top right and bottom), 47 (centre left), 151 (bottom left)
Featherstone & Durber, Sheffield: p. 69 (centre and bottom)
Hemera Technologies Inc.: pp 7 (top right, 8 top left and right, 16, 18, 21, 22, 23, 24, 25, 26, 72, 73 (top right and bottom), 108, 144 (top right)
IKEA: pp. 110–111, 144 (right), 145
J. Allan Cash: p. 131 (centre right)
Jet Mayor: p. 20
Joel Degan: pp. 46 (centre top – Clarissa Mitchell), 54 (centre), 58 and cover (Claire Underwood), 54 (top – Alison Counsell), 64 (students' work, London Guildhall University)
Martyn Chillmaid: pp. 44 (right), 53 (centre left and bottom), 74, 76, 77, 82, 92 (top), 105, 108, 115, 126 (centre left), 139 (bottom), 144 (left)
memphis/Studio Aldo Ballo: p. 117 (top and bottom left); memphis/Studio Azzurro: p. 117 (bottom centre)
Robert Harding Picture Library: pp. 70, 93, 98 (Adam Woolfitt)
Science Pictures Ltd: pp. 78, 130–31 (top), 133 (centre right and bottom), 135 (centre left and bottom right)
Simon Phillips: pp. 31, 34, 36, 43, 59, 60, 61 (centre), 68, 71 (centre and top), 87, 102, 107 (top)
Sony: p 8 (bottom left)
Telegraph Colour Library; p. 9 (bottom – Gone Loco)
Tristram Shepard: pp. 65 (bottom), 107 (centre and bottom), 122–23
ZEFA: pp. 48 (top left – J. Brandenburg), 51 (bottom), 89

The publishers would also like to thank:
British Standards Institute, Maylands Avenue, Hemel Hempstead HP2 4SQ (Tel: 01442 230442)
Craftspace Touring: quotes and images from their catalogue *Recycling: Forms for the Next Century* – Austerity for Posterity on pp. 56, 78
memphis, Milan: images on p. 116
News International: *The Times* article on p. 137
Once a Tree, for their support towards the traditional toys project on p. 74

Contents

Project three: Traditional Toys

Project four: Flat-pack Furniture

Project Suggestions

Introduction

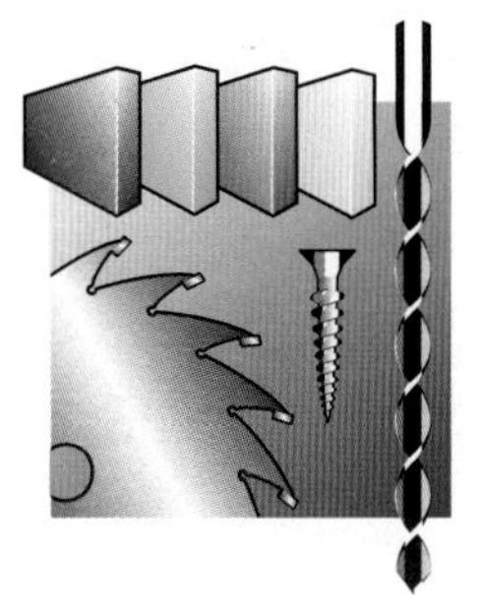

Welcome to* Design & Make It: Resistant Materials. *This book has been written to support you as you work through your GCSE course in Design and Technology. It will help guide you through the important stages of your coursework, and assist you in preparing for the final examination paper.

■ ACTIVITY

Make sure that as part of your design folder you include evidence of having completed a number of short-term focused practical tasks as suggested in the Activity sections.

Long or short?

If you are following a short course, check with your teacher which sections of the book you do not need to cover.

Making it

Whatever your project, remember that the final realisation is particularly important. It is not enough just to hand in your design folder. You must have separate products which you have made. The quality of your final realisation must be as high as possible as it counts for a high proportion of the marks.

During your course you will need to develop technical skills in using resistant materials and equipment. This is something you can't do just by reading a book! The best way is to watch carefully as different techniques and procedures are demonstrated to you, and practise them as often as possible.

How to Use this Book

There are two main ways the book might be used.

1 Follow the four design and make projects in sequence over the whole course, including a selection of the suggested activities (i.e. focused practical tasks). This will ensure complete syllabus coverage. You do not necessarily need to take all of them as far as the production of a finished working product: discuss this with your teacher.
2 Undertake alternative projects to one or more of those provided and refer to those pages which cover the specific areas of knowledge and understanding defined in the examination syllabus and the KS4 National Curriculum.

Contents

Project Guide

The book begins with a coursework guide which summarises the design skills you will need for extended project work. Refer back to these pages throughout the course.

The projects

Four projects are provided. These each contain a mixture of product analysis and development pages and knowledge and understanding pages (e.g. Mechanisms, Ergonomics, Plastics) which include short focused tasks. In each of the projects the development of one possible solution has been used as an ongoing example. You could closely base your own work on this solution, but if you want to achieve higher marks you will need to try to come up with ideas of your own.

Project suggestions

Finally three outline project suggestions are provided. Refer to the Project Guide to help develop your ideas and to ensure you are covering and documenting your coursework in the way the examiners will be looking for.

IN YOUR PROJECT

The 'In Your Project' paragraphs will help you to think about how you could apply the content of the page to your current work.

KEY POINTS

Use the 'Key Points' paragraphs to revise from when preparing for the final examination paper. Three specimen papers are included at the end of the projects.

Beyond GCSE

There are good opportunities for skilled people to work in manufacturing. Another alternative is to train as a product, furniture or jewellery designer, or to specialise in craftwork. Such people need to be flexible, good communicators, willing to work in teams, and to be computer literate.
There are a wide range of further courses and training opportunities available at various levels which you might like to find out more about.

Design Matters

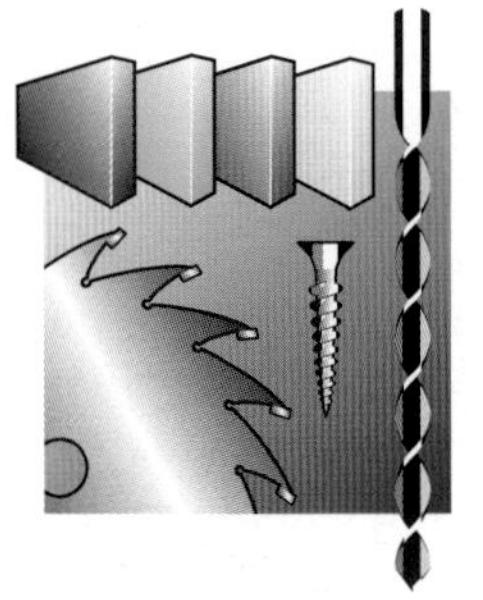

What is Design and Technology, how has it changed, and why is it important?

As you develop your ideas for products you will often need to make important decisions about the social, moral, cultural and environmental impact of your product.

How does Design and Technology Affect our Lives?

Technology helps extend our natural capabilities. For example, it enables us to:

- ▷ travel further and faster
- ▷ send and receive messages across the world in an instant
- ▷ keep us warm in winter and cool in summer.

Designers help make new and existing technologies easy and more pleasant for people to use – they make them look and feel fun and fashionable, logical and safe to use. They also work out how to make them easy to produce in quantity, and cheap to manufacture and sell.

So Design and Technology is about improving people's lives by designing and making the things they need and want. But different people have different needs: what is beneficial to one person can cause a problem for someone else, or create undesirable damage to the environment.

A new design might enable someone to do something quicker, easier and cheaper, but might cause widespread unemployment or urban decay. It could also have a harmful impact on the delicate balance of nature.

As you develop your design ideas you will often need to make important decisions about the social, moral, cultural and environmental impact of your product.

A brief history of resistant materials technology

The first 3D products to be designed were probably arrowheads and simple knives, crudely shaped from pieces of stone. Early ceramic pots and other containers were often elaborately decorated.

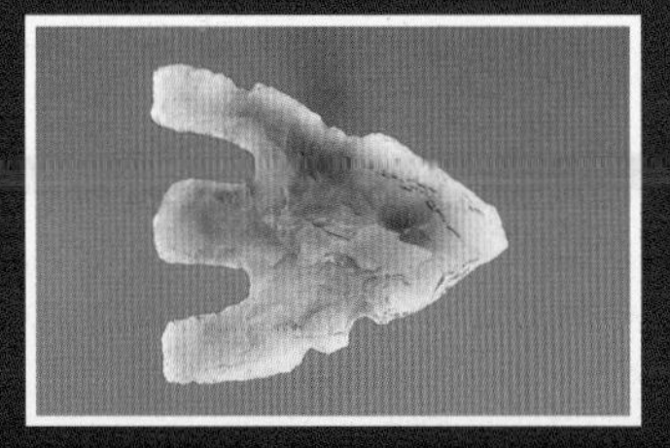

During the Industrial Revolution, engineers and factory owners tended to concentrate on how things could be made to work, and how they could be made in quantity as cheaply as possible. The mass manufacture of products led to much simpler styles which were easier for machines to make.

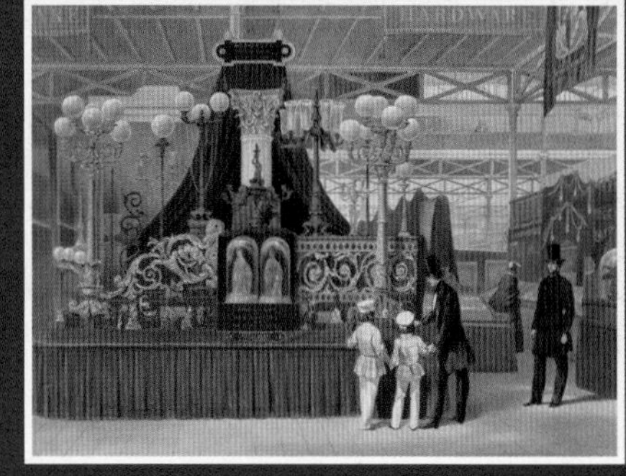

Great Exhibition, 1851.

Environmental Issues

We all need to be aware of the amount and use of energy and resources, as on our earth these are finite. The more we use natural resources without replacing them, the fewer there are for future generations. Many products use plastics derived from petro-chemicals that use non-replaceable energy and contribute to global warming.

Social Needs

Good design can help bring people together. Designers need to be careful about creating products that might have the effect of isolating someone, or making them more vulnerable to crime in some way.

Cultural Awareness

People from different cultures think and behave in different ways. What is acceptable to one culture may be confusing or insulting to another.
For example, in Britain most families use kettles to boil drinking water. This is not the case in France where many use a pan. Colours and certain shapes can have very different meanings across the world.

Moral Issues

Sometimes designers are asked to develop products that can cause harm to people or animals. Would you be willing to create a product that could hurt someone or could also be used for criminal activity?

At the start of the 20th century the designers at the Bauhaus in Germany tried hard to make machine-made products look good. Their design principles are still copied today.

Sometimes historical events dictate style. For example many products in the 1920s and 1930s had an Egyptian look after the discovery of Tutankhamen's tomb in 1922.

In the post-war 1950s and 1960s there was a consumer boom which led to the design of 'throwaway' products in which fashion became more important than function. Late 1960s products were influenced by the earlier invention of the transistor, space exploration and 'psychedelic' colours and patterns.

Industrial Matters

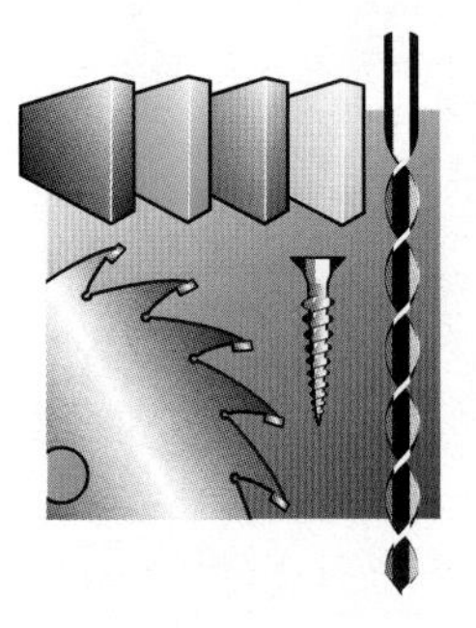

Good design involves creating something that works well and is satisfying to use. But to be successful a product also needs to be commercially viable.

Professional Designers

Different designers specialise in creating different sorts of products. Professional designers usually specialise in different areas. For example:

- ▷ 3D Design
- ▷ 2D Design
- ▷ Fashion and Textiles
- ▷ Architecture.

This book is particularly concerned with designing and making 3D products using resistant materials, such as woods, metals and plastics.

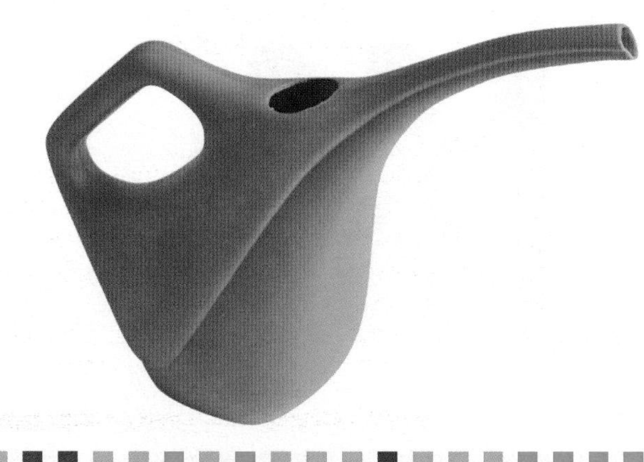

What is a Resistant Materials Product?

We are surrounded by many three-dimensional products made from resistant materials that we use every day, such as:

- ▷ furniture – tables, chairs, cupboards, etc.
- ▷ electrical appliances – kettles, power-drills, hair-dryers, personal stereos, etc.
- ▷ jewellery and other decorative items
- ▷ toys, games, sports equipment, etc.
- ▷ storage containers
- ▷ table-ware
- ▷ lighting devices

What others can you think of?

To design and make a successful 3D product you need to know what people need and want, and how materials and production technologies can be used to create it. You also need some skills in designing and making.

The theme of the last quarter of the 20th century has been miniaturisation, led by substantial developments in micro-electronics and new materials technologies. Product design is now often led by sophisticated marketing of designer labels and global brands.

Design for Profit

Products are designed and made to make life easier and more enjoyable, or to make a task or activity more efficient. However, along the way the people who create these products need to make a profit. The designer needs to do more than satisfy the needs of the market, and to consider the sorts of issues described on the previous page. They must also take into account the needs of the clients, manufacturers and retailers. The aim is to design and make products that are successful from everyone's point of view.

Designers:

- agree a brief with a client
- keep a notebook or log of all work done with dates so that time spent can be justified at the end of the project
- check that an identified need is real by examining the market for the product
- keep users' needs in mind at all times
- check existing ideas. Many designers re-style existing products to meet new markets because of changes in fashion, age, environment, materials, new technologies, etc.
- consider social, environmental and moral implications
- consider legal requirements
- set limits to the project to guide its development (design specification)
- produce workable ideas based on a thorough understanding of the brief
- design safe solutions
- suggest materials and production techniques after considering how many products are to be made
- produce working drawings for manufacturers to follow.

Retailers:

- need to make a profit on the products sold
- consider the market for the product
- give consumers what they want, when they want it, at an acceptable price
- take account of consumers' legal rights
- take consumer complaints seriously
- continually review new products
- put in place a system to review and replace stock levels.

Manufacturers need to:

- make a profit on the products produced
- agree and set making limits for the product (manufacturing specification)
- develop marketing strategies
- understand and use appropriate production systems
- reduce parts and assembly time
- reduce labour and material costs
- apply safe working procedures to make safe products
- test products against specifications before distribution
- produce consistent results (quality assurance) by using quality control procedures
- understand and use product distribution systems
- be aware of legislation and consumer rights
- assume legal responsibility for product problems or failures.

Consumers/users expect the product to:

- do the job it was designed for
- give pleasure in use
- have aesthetic appeal
- be safe for its purpose
- be of acceptable quality
- last for a reasonable lifetime
- offer value for money.

Clients:

- identify a need or opportunity and tell a designer what they want a product to do and who it's for (the brief)
- consider the possible market for the idea
- organise people, time and resources and raise finance for the project.

During the first decade of the 21st century developments in electronic communication technologies will continue to develop rapidly in new and exciting ways.

The future of product design lies in a further convergence of electronics and communication devices, new 'smart' materials that respond to their environments, and personalised target marketing.

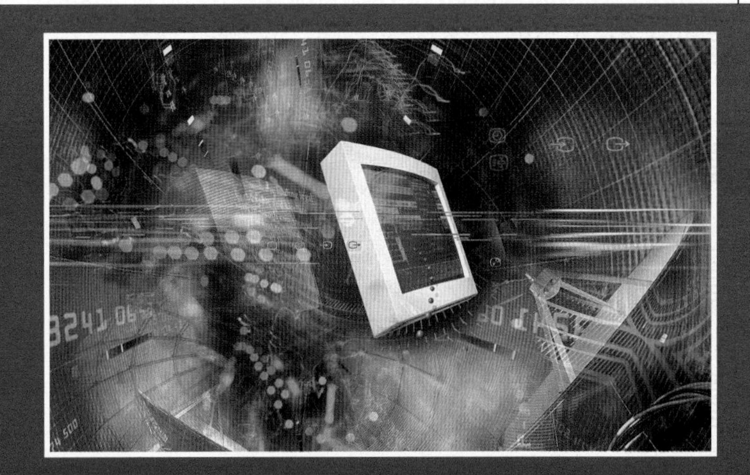

Using ICT in Resistant Materials Technology

ICT (Information and Communication Technology) is widely used in the design and production of products, as you will discover. You can considerably enhance your GCSE coursework with the effective use of ICT.

Using ICT in your Work

To gain credit for using ICT you need to know when it is best to use a computer to help with your work. Sometimes it is easier to use ICT to help with parts of your coursework than to do it another way. On other occasions it can be far easier to write some notes on a piece of paper than use a computer – this saves you time and helps you to do the job more effectively.

The following are some ideas showing you how using ICT could enhance your coursework. Some can be used at more than one stage. You do not have to use all of them!

Identifying the Problem

The **Internet** could be used to search manufacturers' and retailers' web sites for new products, indicating new product trends.

Project Planning

A time chart can be produced showing the duration of the project and what you hope to achieve at each stage using a **word processor** or **DTP** program. Some programs allow you to produce a Gantt chart (see page 129).

Investigation

- A questionnaire can be produced using a **word processor** or **DTP**. Results from a survey can be presented using a **spreadsheet** as a variety of graphs and charts.
- Use a **digital camera** to record visits and existing products
- The **Internet** can be used to perform literature searches and to communicate with other people around the world via **e-mail**.

Search engines

To help you find the information you need on the Internet you can use a search engine. A search engine is a web site that allows you to type in keywords for a specific subject. It then scans the Internet for web sites that match what you are looking for. Here are the addresses of some popular search engines:

www.excite.co.uk
www.yahoo.com
www.netscape.com
www.hotbot.lycos.co.uk
www.msn.co.uk
www.searchtheweb.com

E-mail

E-mail is a fast method of communicating with other people around the world. Text, photographs and computer files can be attached and shared. Some web-sites have e-mail addresses – you could try to contact experts to see if they could help with your coursework. It is important to be as specific as you possibly can, as these experts may be very busy people.

Specification

A design or product specification can be written with a **word processor**. Visual images of the product, diagrams and other illustrations could also be added. Information can be easily modified at a later date.

Developing Initial Ideas

Ideas for your product could be produced using a **graphics** program, **DTP** or **3D design** package. Colour variations can be applied to product drawings to test a design on its intended market before production.

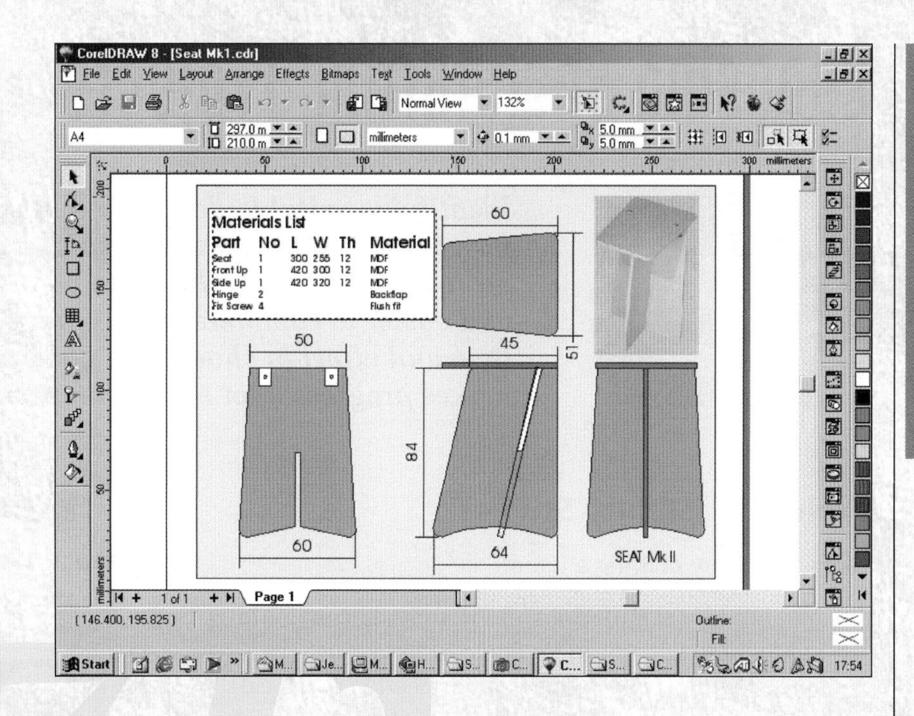

Likely costs of new products can be modelled using a **spreadsheet**. Different component costs can be modelled quickly and easily allowing you to see the consequences of your design ideas.

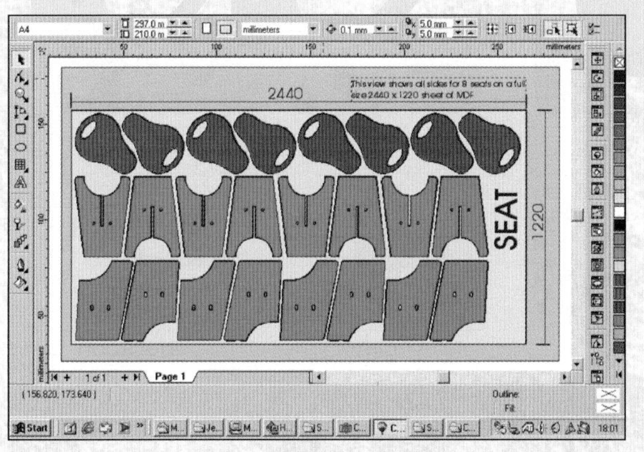

The seat shown above was designed to carry and store flat to save space. It was sketched first to get the general concept right and then finalised using CAD. The intention was to use an industrial flatbed CNC router to cut out all the parts. Once the machine is set many sheets can be cut with repeated accuracy.

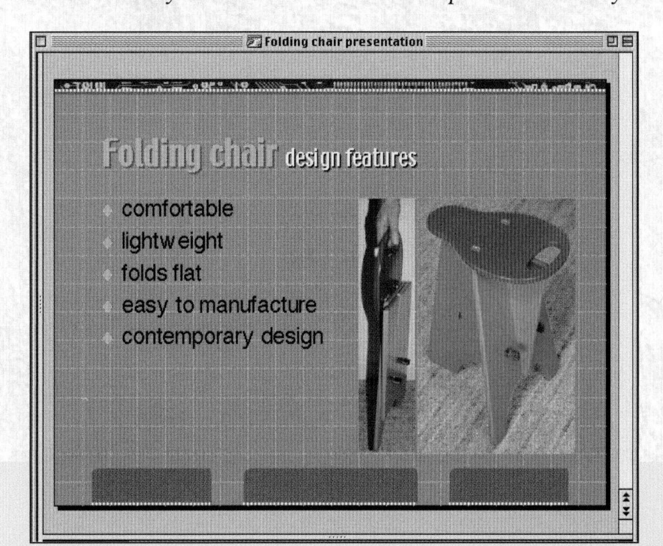

Final Ideas and Production

A document showing the specification, images, production method and components can be **word processed**.

Parts lists and the costs of materials can be calculated and displayed using **spreadsheets**.

A detailed flow production diagram could be produced using a **DTP** program. Images could also be added to show important stages.

Digital images can be used in the production plan as a guide to show how the product should be assembled or to indicate its colour.

Pre-programmed **CAM** equipment could be used to replicate manufacture (see page 136).

Packaging nets can be produced using **graphics package** or by using templates, e.g. **www.dtonline.org**. **Scanned** or **digital photographs** can then be placed within the templates to produce instant packaging designs.

Project Presentation

Use **graphics packages** to prepare text and visual material for presentation panels. Charts showing numerical data can be quickly produced using a **spreadsheet**.

Use a **presentation** package, such as *PowerPoint* to communicate the main features of your design.

Choosing and Starting Projects

Identifying suitable design and make projects for yourself is not easy. A carefully chosen project is much more likely to be interesting and easier to complete successfully. Investing time and effort in choosing a good project makes progress a lot easier later on.

Project Feasibility Studies

Make a start by making a list of:

- ▷ potential local situations/environments you could visit where you could do some research into the sort of things people there might need (e.g. a local playgroup, a small business, a hospital or sports centre, etc.)
- ▷ people you know outside school who might be able to help by providing information, access and/or advice.

The next stage is to get up and get going. Arrange to visit some of the situations you've listed. Choose the ones which you would be interested in finding out some more about. Make contact with the people you know, and get them interested in helping you. Tell them about your D&T course, and your project.

For each possible situation you should:

- ▷ visit the situation or environment
- ▷ make initial contact with those whose help you will need.

With a bit of luck, after you've done the above you should have a number of ideas for possible projects.

Try to identify what the possible outcomes of your projects might be – not what the final design would be, but the form your final realisation might take, e.g. a working object, a scale model, a series of plans and elevations, etc. Think carefully about the following:

- ▷ Might it be expensive or difficult to make?
- ▷ Do you have access to the tools and materials which would be required?
- ▷ Will you be able to find out how it could be manufactured?
- ▷ Does the success of the project depend on important information you might not be able to get in the time available?
- ▷ Are there good opportunities for you to use ICT?

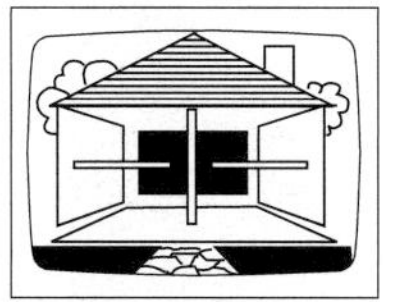
the home

energy

the natural environment

the high street

transport

communications

clothing

leisure

security

food

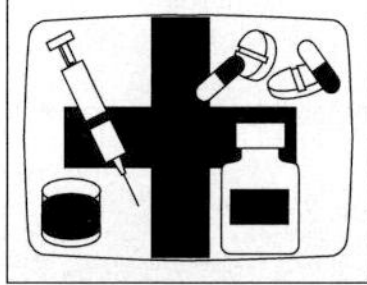
health

education

starting points

There are a number of different ways in which you might start a project. Your teacher may have:

- told you exactly what you are required to design
- given you a range of possible design tasks for you to choose from
- left it up to you to suggest a possible project.

If you have been given a specific task to complete you can go straight on to page 14.

If you are about to follow one of the main units in this book, you should go straight to the first page of the task.

If you have been given a number of possible tasks to choose from you should go straight to the section on page 13 entitled 'Making your Choice'.

However, if you need to begin by making some decisions about which will be best task for you, then the first stage is to undertake some project feasibility studies as described on this page.

Making Your Choice

For each of your possible projects work through pages 14 and 15 (Project Investigation) and try planning out a programme of research.

Look back over your starting questions and sources of information:

- ▷ Could you only think up one or two areas for further research?
- ▷ Did you find it difficult identifying a range of sources of information?

If this has been the case, then maybe it is not going to prove to be a very worthwhile project.

Ideally, what you're looking for is a project which:

- ▷ is for a nearby situation you can easily use for research and testing
- ▷ you can get some good expert advice about.
- ▷ shows a good use of ICT.

It is also important that your expected outcome:

- ▷ will make it possible for you to make and test a prototype
- ▷ will not be too difficult to finally realise.

Finally, one of the most important things is that you feel interested and enthusiastic about the project!

don't forget...

Don't forget to record all your thoughts and ideas about these initial stages of choosing and starting your project.

in my design folder

- ✓ *My project is to design a...*
- ✓ *I am particularly interested in...*
- ✓ *I have made a very good contact with...*
- ✓ *My prototype can be tested by...*
- ✓ *My final outcome will include...*
- ✓ *I could use ICT to...*

Project Investigation

You will need to find out as much as you can about the people and the situation you are designing for. To do this you will need to identify a number of different sources of information to use for your research.

Using a spreadsheet is an excellent way of presenting data

Starting Questions

Make a list of questions you will have to find answers to.

You should find the following prompts useful:

Why...?
When...?
Where...?
What...?
How many...?
How often...?
How much...?

Sources of Information

Next, carefully consider and write down the potential sources of information you might be able to use in order to discover the answers to your starting questions.

Work through the research methods on the next page. Be sure to give specific answers as far as possible (i.e. name names).

Across your research you will need to aim to obtain overall a mixture of:

- factual information: e.g. size, shape, weight, cost, speed, etc.
- information which will be a matter of opinion: i.e. what people think and feel about things, their likes and dislikes, what they find important, pleasing, frustrating, etc.

don't forget...

Write down what you need to find out more about, and how you could obtain the information.

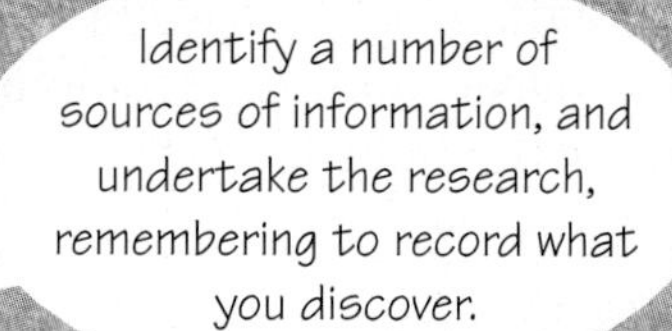

Make sure your research work is clearly and attractively presented.

You need to identify a number of sources of information (e.g. user research, existing solutions, expert opinion, information search). The wider the range of methods you use, the more marks you will get.

in my design folder

- ✓ The key things I need to find out about are...
- ✓ The research methods I am going to use are...
- ✓ I will be talking to the following people about my project...
- ✓ I will need to have it all completed by...
- ✓ I will use ICT to...

Research Methods

User Research

Which people could you observe and consult who are directly involved in the situation? To what extent do you consider that you will be able to find out about:

- the things they do
- the way in which they do them.

As well as asking individuals, you could also undertake a small survey or questionnaire.

User Trips

How can you record your own impressions of the situation? Are there any relevant activities you could try out for yourself to gain first-hand experience? Do you have any recollections of any previous similar experiences you have had?

Site Study

In what ways could you document the environment in which the problem is? Which of the following will be relevant?

- Historical and geographical factors.
- Sociological, economic, political information.
- Location.
- Layout, facilities.
- Sizes and spaces.
- Atmosphere – light, colour, texture.
- The surrounding environment.

Similar Situations

Do you know of any other comparative circumstances in which people are in similar situations, and which might help provide insight and ideas?

Expert Opinion

Are there any people you know of who could give you expert professional advice on any aspects of the situation? If you don't know immediately of anyone, how might you set about finding somebody?

Information Search

Has any information about the situation, or a similar situation, been documented already in books, magazines, TV programmes, the Internet, or CD-ROM?
If you don't already know that such information exists, where could you go to look for it? Don't forget to consider the possibility of using information stored on a computer database.

In Conclusion

When most of your investigation work has been completed you will need to draw a series of conclusions from what you have discovered. What have you learnt about the following things:

▷ What sort of people are likely to be using the product?
▷ Where and when will they be using it?
▷ What particular features will it need to have?
▷ How many should be made?

Of all the research methods, user-research tends to be the most effective and useful, so you are highly recommended to include some in your investigation. Some form of personal contact is essential to a really successful project.

It is also highly advisable to conduct some form of questionnaire. If you have not done one to submit as part of your coursework, make sure that you will have the opportunity to do so this time.

It isn't necessary to use all the research methods in any one project, but you certainly must show that you have tried a selection of them.

in my design folder

✓ From my research I found out...
✓ I have discovered that...
✓ My conclusions are...
✓ I have kept my research relevant by...
✓ I found ICT helpful when...

From Design Specification to Product Specification

A design specification is a series of statements that describe the possibilities and restrictions of the product. A product specification includes details about the features and appearance of the final design, together with its materials, components and manufacturing processes.

Writing a Design Specification

The **design specification** is a very important document. It describes the things about the design which are fixed and also defines the things which you are free to change.

The conclusions from your research should form the basis of your design specification. For example, if in your conclusions you wrote:

'From the measurements I made of a number of people's arm lengths, I discovered that the best size for an arm-rest would be between 250 mm and 400 mm long.'

In the specification you would simply write:

'The arm-rest should be between 250 mm and 400 mm in length.'

The contents of the specification will vary according to the particular product you are designing, but on the next page is a checklist of aspects to consider. Don't be surprised if the specification is quite lengthy. It could easily contain 20 or more statements.

Fixing It

Some statements in the specification will be very specific, e.g.:*'The toy must be red.'*

Other statements may be very open ended, e.g.: *'The toy can be any shape or size.'*

Most will come somewhere in between, e.g.: *'The toy should be based on a vehicle of some sort and be mechanically or electronically powered.'*

In this way the statements make it clear what is already fixed (e.g. the colour), and what development is required through experimentation, testing, and modification (e.g. shape, size, vehicle-type and method of propulsion).

Writing a Product Specification

After you have fully developed your product you will need to write a final more detailed **product specification**. This time the precise statements about the materials, components and manufacturing processes will help ensure that the manufacturer is able to make a repeatable, consistent product.

Your final product will need to be evaluated against your design specification to see how closely you have been able to meet its requirements, and against your product specification to see if you have made it correctly.

don't forget...

You might find it helpful to start to rough out the design specification first, and then tackle the conclusions to your research. Working backwards, a sentence in your conclusion might need to read:

'From my survey, I discovered that young children are particularly attracted by bright primary colours.'

It's a good idea to use a word processor to write the specification. After you've written the design specification new information may come to light. If it will improve the final product, you can always change any of the statements.

Make sure you include as much numerical data as possible in your design specification. Try to provide data for anything which can be measured, such as size, weight, quantity, time and temperature.

Specification Checklist

The following checklist is for general guidance. Not all topics will apply to your project. You may need to explore some of these topics further during your product development.

Use a word processor to draft and finalise your design specification.

Use and performance

Write down the main purpose of the product – what it is intended to do. Also define any other particular requirements, such as speed of operation, special features, accessories, etc. Ergonomic information is important here.

Size and weight

The minimum and maximum size and weight will be influenced by things such as the components needed and the situation the product will be used and kept in.

Generally the smaller and lighter something is the less material it will use, reducing the production costs. Smaller items can be more difficult to make, however, increasing the production costs.

Appearance

What shapes, colours and textures will be most suitable for the type of person who is likely to use the product? Remember that different people like different things.

These decisions will have an important influence on the materials and manufacturing processes, and are also crucial to ensure final sales.

Safety

A product needs to conform to all the relevant safety standards.

- Which of them will apply to your design?
- How might the product be mis-used in a potentially dangerous way?
- What warning instructions and labels need to be provided?

Conforming to the regulations can increase production costs significantly, but is an area that cannot be compromised.

Manufacturing cost

This is concerned with establishing the maximum total manufacturing cost which will allow the product to be sold at a price the consumer or client is likely to pay.

The specification needs to include details of:

- the total number of units likely to be made
- the rate of production and, if appropriate
- the size of batches.

Maintenance

Products which are virtually maintenance free are more expensive to produce.

- How frequently will different parts of the product need to be maintained?

Life expectancy

The durability of the product has a great influence on the quantity of materials and components and the manufacturing process which will need to be used.

How long should the product remain in working order, providing it is used with reasonable care?

Environmental requirements

In your specification you will need to take into account how your product can be made in the most environmentally friendly way. You might decide to:

- specify maximum amounts of some materials
- avoid a particular material because it can't be easily recycled
- state the use of a specific manufacturing process because it consumes less energy.

Other areas

Other statements you might need to make might cover special requirements such as transportation and packaging.

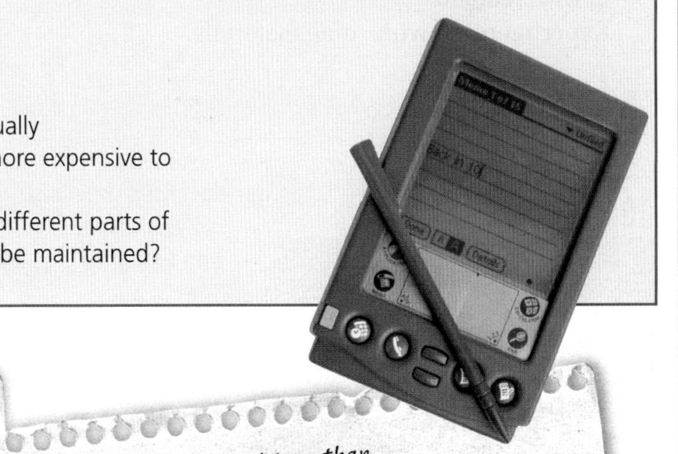

in my design folder

- ✓ My design will need to...
- ✓ The requirements of the people who will use it are...
- ✓ It will also need to do the following...
- ✓ It will be no larger than...
- ✓ It will be no smaller than...
- ✓ Its maximum weight can be...
- ✓ It should not be lighter than...
- ✓ The shapes, colours and textures should...
- ✓ The design will need to conform to the following safety requirements...
- ✓ The number to be printed or made is...
- ✓ The following parts of the product should be easily replaceable...
- ✓ To reduce wastage and pollution it will be necessary to ensure that...

Generating and Developing Ideas (1)

When you start designing you need lots of ideas – as many as possible, however crazy they might seem. Then you need to start to narrow things down a bit by working in more detail and evaluating what you are doing.

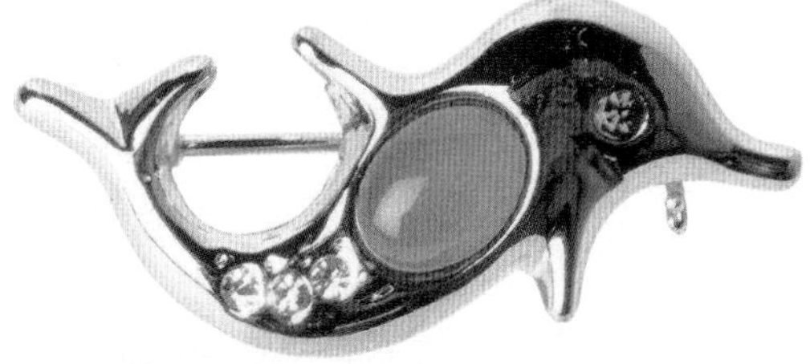

First Thoughts

Start by exploring possibilities at a very general level. Spend time doing some of the following:

- ▷ Brainstorming, using key words and phrases or questions which relate to the problem.
- ▷ Completing spider-diagrams which map out a series of ideas.
- ▷ Using random word or object-association to spark off new directions.
- ▷ Thinking up some good analogies to the situation (i.e. what is it like?).
- ▷ Work from an existing solution by changing some of the elements.
- ▷ Experimenting with some materials.

Continue doing this until you have at least two or three possible approaches to consider. Make sure they are all completely different, and not just a variation on one idea.

Go back to your design specification. Which of your approaches are closest to the statements you made? Make a decision about which idea to take further, and write down the reasons for your choice.

ICT 

Wherever possible consider using a computer to experiment with your ideas, and to analyse and present your findings.

As you work through this section it is important to remember the following sequence when considering potential solutions:

- record a number of different possibilities
- consider and evaluate each idea
- select one approach as the best course of action, stating why.

There are lots of different drawing techniques which you can use to help you explore and develop your ideas, such as plans, elevations, sections, axonometrics, perspectives, etc.
Try to use as rich a mixture of them as possible. At this stage they should really be 'rough', rather than 'formal' (i.e. drawn with a ruler). Colour is most useful for highlighting interesting ideas.

don't forget...

As usual, it is essential to record all your ideas and thoughts.

Much of your work, particularly early on, will be in the form of notes. These need to be neat enough for the examiner to be able to read.

Drawings on their own do not reveal very well what you had in mind, or whether you thought it was a good idea or not. Words on their own suggest that you are not thinking visually enough. Aim to use both sketches and words.

Second Thoughts

Working on paper, begin to develop your ideas in more detail. Remember to use a range of drawing techniques, such as plans, elevations, 3D sketches, as well as words and numbers to help you model your ideas.

Getting ideas down quickly

It is a real advantage to be able to get your ideas down on paper very quickly. Try sketching your ideas without using colour or a ruler. It doesn't matter if your sketches don't look accurate or very realistic at this stage. Get lots of ideas down on paper so that you can discuss them with other people.

Being able to sketch quickly can be learned, just like you learned to write words. At first you will need lots of practise and probably some help. Try copying or tracing good sketches done by others so that you get a feel for it.

Theme for a Dream

Have some fun with your ideas. You don't need to stick to traditional shapes, colours, patterns or textures. You could use some sort of theme for the look of your products, for example:

▷ Space travel, space monster
▷ Mexican patterns, musical instruments
▷ Architecture from foreign lands
▷ Computer games characters
▷ Festive activities
▷ Summer holidays
▷ Animal or geometric shapes

If you use this technique for early ideas you will be able to come up with lots of them very easily. This is one of the ways designers in industry come up with fresh ideas time after time.

Make notes on your sketches if you want to remember a small detail or indicate something you can't sketch very well.

Remember your sketches tell a story as they show how you have progressed with your ideas. Some of the story might save you time if you use words. For example you might want to say 'glue and screw here' or 'use hinges' or 'fix it together in this order'. Use the quickest methods you can to get to an end result.

Generating and Developing Ideas (2)

CAD-CAM can be extremely useful at this stage. Work towards making at least one prototype to test out some specific features of your design. Record the results and continue to refine your ideas as much as you can. Sorting out the final details often requires lots of ideas too.

CAD / CAM

Computer Aided Design (CAD) and **Computer Aided Manufacture (CAM)** are terms used for a range of different ICT applications that are used to help in the process of designing and making products.

CAD is a computer-aided system for creating, modifying and communicating ideas for a product or components of a product.

CAM is a broad term used when several manufacturing processes are carried out at one time aided by a computer. These may include process control, planning, monitoring and controlling production.

The cups shown top right were designed and shaded (rendered) using a CAD system. The cups above are real models which were made to test the size and feel of the cups before manufacture. They are made from a plastic resin which is easy to machine and spray paint. The handles were stuck on after machining.

CAM or CNC systems may be used for rapid prototyping or modelling in 'soft' materials, the manufacture of small parts and moulds, and the production of templates, stencils or printouts on paper, card, vinyl, etc.

You may not have CAD-CAM or CNC systems in your workshops but you can still indicate how batch production of your product might make use of CNC machines like lathes, milling machines, flatbed routers, spindle moulders, injection and extrusion machines, etc.

CIM

Computer Integrated Manufacture is extensively used by industry. This is where computers are used to control the whole production process from materials input and handling, to pick and placement, to machining to construction and possible packaging.

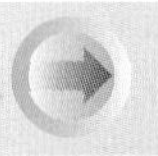

ICT

Wherever possible consider using a CAD program. Designing on-screen happens very quickly and little evidence of change is seen compared with drawing on paper. You need to develop ways of recording your thinking and collect evidence of any developments.

- Make sure you print out the various stages you work through, or keep a copy on disk.
- Where CAM/CNC is used you must record all your programming and evidence of machine set-ups in your project folder.

As you develop your ideas, make sure you are considering the following:

- Design – aesthetics, ergonomics, marketing potential, etc.
- User requirements – functions and features.
- Technical viability – if it could be made.
- Manufacturing potential – how it could be made in quantity.
- Environmental concerns – if it can be reused, recycled, etc.?

Models are simplified versions of intended products. Use words, numbers, drawings and 3D representations of your ideas to help you develop and evaluate your designs as they progress.

Planning and Making Prototypes

At some stage you will be in a position to bring your ideas into sharper focus by making some form of mock-up or prototype. Think carefully about exactly what aspect of your idea you want to test out and about the sort of model which will be most appropriate.

Whatever the form of your final outcome, the prototype might need to be:

- two-dimensional
- three-dimensional
- at a different scale
- made using different materials.

Try to devise some objective tests to carry out on your prototype involving measuring something. Don't rely just on people's opinions. Write up the circumstances in which the tests were undertaken, and record your results.

Write down some clear statements about:

- what you wanted the prototype to test
- the experimental conditions
- what you discovered
- what decisions you took about your design as a result.

Following your first prototype you may decide to modify it in some way and test it again, or maybe make a second, improved version from scratch. Make sure you keep all the prototypes you make, and ideally take photographs of them being tested perhaps using a digital camera.

Sometimes you will need to go back to review the decisions you made earlier, and on other occasions you may need to jump ahead for a while to explore new directions or to focus down on a particular detail. Make sure you have worked at both a general and a detailed level.

in my design folder

- ✓ *I chose this idea because...*
- ✓ *I developed this aspect of my design by considering...*
- ✓ *To evaluate my ideas I decided to make a prototype which...*
- ✓ *The way I tested my prototype was to...*
- ✓ *What I discovered was...*
- ✓ *As a result I decided to change...*
- ✓ *I used ICT to...*

Planning the Making and the Manufacturing

The final realisation is very important. It presents your proposed design solution rather than the process you used to develop it. Careful planning is essential. You will also need to be able to explain how your product could be manufactured in quantity.

How many?
What you have designed should be suitable for manufacture. You should discuss with your teacher how many items you should attempt to make. This is likely to depend on the complexity of your design and the materials and facilities available in your workshops. It may be that you only make one item, but also provide a clear account of how a quantity of them could be manufactured.

keeping a record

Write up a diary record of the progress you made while making. Try to include references to:

- things you did to ensure safety
- the appropriate use of materials
- minimising wastage
- choosing tools
- practising making first
- checking that what you are making is accurate enough to work
- asking experts (including teachers) for advice explaining why you had to change your original plan for making.

A Plan of Action

Before you start planning you will need to ensure that you have an orthographic drawing of your design (see page 38). This will need to include all dimensions and details of the materials to be used.

Ideally there should also be written and drawn instructions which would enable someone else to make up the design from your plans.

Next work out a production flow chart as follows.

1 List the order in which you will make the main parts of the product. Include as much detail as you can

2 Divide the list up into a number of main stages, e.g. gathering materials and components, preparing (i.e. marking out, cutting), assembling, finishing.

3 Identify series of operations which might be done in parallel.

4 Indicate the time scale involved on an hourly, daily and weekly basis.

Consider the use of templates and jigs to help speed things up. Other possibilities include the use of moulds or setting up a simple CAM system to produce identical components (see page 136).

don't forget...

You may find you have to change your plans as you go. There is nothing wrong with doing this, but you should explain why you have had to adjust your schedule, and show that you have considered the likely effect of the later stages of production.

Try using the just in time technique described on page 129.

Quality Counts

As your making proceeds you will need to check frequently that your work is of acceptable quality. How accurately will you need to work? What tolerances will be acceptable (see page 138)? How can you judge the quality of the finish?

If you are making a number of identical items you should try and work out ways of checking the quality through a sampling process (see page 139).

Making

While you are in the process of making you must ensure that the tools and materials you are using are the correct ones. Pay particular attention to safety instructions and guidelines.

Try to ensure that you have a finished item at the end, even if it involves simplifying what you do.

Aim to produce something which is made and finished as accurately as possible. If necessary you may need to develop and practise certain skills beforehand.

Manufacturing matters

Try asking the following questions about the way your design might be made in quantity:

- What work operation is being carried out, and why? What alternatives might there be?
- Where is the operation done, and why? Where else might it be carried out?
- When is it done, and why? When else might it be undertaken?
- Who carries it out, and why? Who else might do it?
- How is it undertaken, and why. How else might it be done?

Remember that manufacturing is not just about making things. It is also about making them better by making them:

- simpler
- quicker
- cheaper
- more efficient
- less damaging to the environment.

Planning for Manufacture

Try to explain how your product would be manufactured in quantity. Work through the following stages:

1. Determine which type of production will be most suitable, depending on the number to be made.
2. Break up the production process into its major parts and identify the various sub-assemblies.
3. Consider where jigs, templates and moulding processes could be used. Where could 2D or 3D CAM be effectively used?
4. Make a list of the total number of components and volume of raw material needed for the production run.
5. Identify which parts will be made by the company and which will need to be bought in ready-made from outside suppliers.
6. Draw up a production schedule which details the manufacturing process to ensure that the materials and components will be available exactly where and when needed. How should the workforce and workspace be arranged?
7. Decide how quality control systems will be applied to produce the degree of accuracy required.
8. Determine health and safety issues and plan to minimise risks.
9. Calculate the manufacturing cost of the product.
10. Review the design of the product and manufacturing process to see if costs can be reduced

More information on all these topics can be found on pages 124 to 141.

in my design folder

- ✓ *I planned the following sequence of making...*
- ✓ *I had to change my plan to account for...*
- ✓ *I used the following equipment and processes...*
- ✓ *I paid particular attention to safety by...*
- ✓ *I monitored the quality of my product by...*
- ✓ *My product would be manufactured in the following way...*

Remember to use a wide range of graphic techniques to help plan and explain your making.

Don't forget that there is also a high proportion of marks for demonstrating skill and accuracy, overcoming difficulties and working safely during the making.

What needs to be done by:
- next month
- next week
- next lesson
- the end of this lesson?

Testing and Evaluation

You will need to find out how successful your final design solution is. How well does it match the design specification? How well have you worked? What would you do differently if you had another chance?

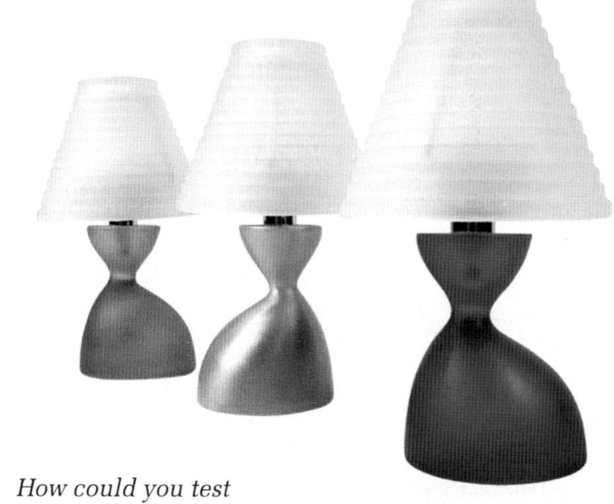

How could you test these products?

As you work through your project you will regularly carry out testing and evaluation. For example:

- analysing and evaluating the research material you collected
- evaluating and testing carried out on existing products
- evaluating initial sketch ideas or samples and models in order to make the right decisions about which to develop further
- assessing the quality of your making as you go along.

Last of all, you must test and evaluate your final solution.

Testing the Final Solution

To find out how successful your design is you will need to test it out. Some of the ways in which you might do this are by:

- trying it out yourself
- asking other people to use it
- asking experts what they think about it.

As well as recording people's thoughts, observations and opinions, try to obtain some data: how many times it worked, over what periods of time, within what performance limits, etc?

To help you decide what to test, you should look back to the statements in your design specification and focus on the most important ones. If for example the specification stated that a three-year-old child must be able to operate it, try and find out if they can. If it must be a colour which would appeal to young children, devise a way of finding out what age ranges it does appeal to.

You need to provide evidence to show that you have tested your final design out in some way. Try to ensure that your findings relate directly to the statements in your original specification. Include as much information and detail as you can.

don't forget...

Final Evaluation

There are two things you need to discuss in the final evaluation: the quality of the product you have designed, and the process you went through while designing it.

The product

How successful is your final design? Comment on things like:

- ▷ how it compares with your original intentions as stated in your design specification
- ▷ how well it solves the original problem
- ▷ the materials you used
- ▷ what it looks like
- ▷ how well it works
- ▷ what a potential user said
- ▷ what experts said
- ▷ whether it could be manufactured cheaply enough in quantity to make a profit
- ▷ the effective use of ICT to assist reproduction or manufacture
- ▷ the extent to which it meets the requirements of the client, manufacturer and the retailer
- ▷ the ways in which it could be improved.

Justify your evaluation by including references to what happened when you tested it.

The process

How well have you worked? Imagine you suddenly had more time, or were able to start again, and consider:

- ▷ Which aspects of your investigation, design development work and making would you try to improve, or approach in a different way?
- ▷ What did you leave to the last minute, or spend too much time on?
- ▷ Which parts are you most pleased with, and why?
- ▷ How well did you make the final realisation?
- ▷ How effective was your use of ICT? How did it enhance your work?

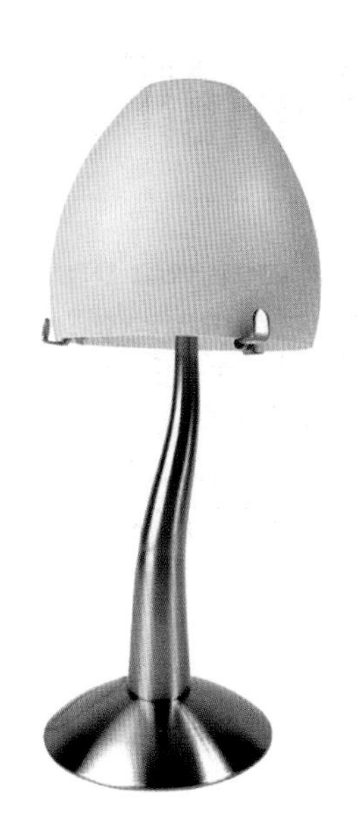

If you had more time:

- what aspects of the product would you try to improve? (refer to your evaluation if you can).
- how would you improve the way you had researched, developed, planned and evaluated your working process?

in my design folder

What do you think you have learnt through doing the project?

- ✓ *Comparison of my final product specification with my design specification showed that...*
- ✓ *The people I showed my ideas (drawings and final product) to said...*
- ✓ *I was able to try my design out by...*
- ✓ *I discovered that...*
- ✓ *I could improve it by...*
- ✓ *I didn't do enough research into...*
- ✓ *I spent too long on...*
- ✓ *I should have spent more time on...*
- ✓ *The best aspect is...*
- ✓ *I have learnt a lot about...*

Project Presentation

The way you present your project work is extremely important. Remember you won't be there to explain it all when it's being assessed! You need to make it as easy as possible for an examiner to see and understand what you have done.

Telling the Story

All your investigation and development work needs to be handed in at the end, as well as what you have made. Your design folder needs to tell the story of the project. Each section should lead on from the next, and show clearly what happened next, and explain why. Section titles and individual page titles can help considerably.

There is no single way in which you must present your work, but the following suggestions are all highly recommended:

- Securely bind all the pages together in some way. Use staples or treasury tags. There is no need to buy an expensive folder.
- Add a cover with a title and an appropriate illustration.
- Make it clear which the main sections are.
- Add titles or running headings to each sheet to indicate what aspect of the design you were considering at that particular point in the project.

Remember to include evidence of ICT work and other Key Skills. Carefully check through your folder and correct any spelling and punctuation mistakes.

don't forget...

Presentation is something you need to be thinking about throughout your project work.

Presenting your Design Project Sheets

- Always work on standard-size paper, either A3 or A4.
- Aim to have a mixture of written text and visual illustration on each sheet.
- You might like to design a special border to use on each sheet.
- Include as many different types of illustration as possible.
- When using photographs, use a small amount of adhesive applied evenly all the way around the edge to secure them to your folder sheet.
- Think carefully about the lettering for titles, and don't just put them anywhere and anyhow. Try to choose a height and width of lettering which will be well balanced on the whole page. If the title is too big or boldly coloured it may dominate the sheet. If it is thin or light it might not be noticed.

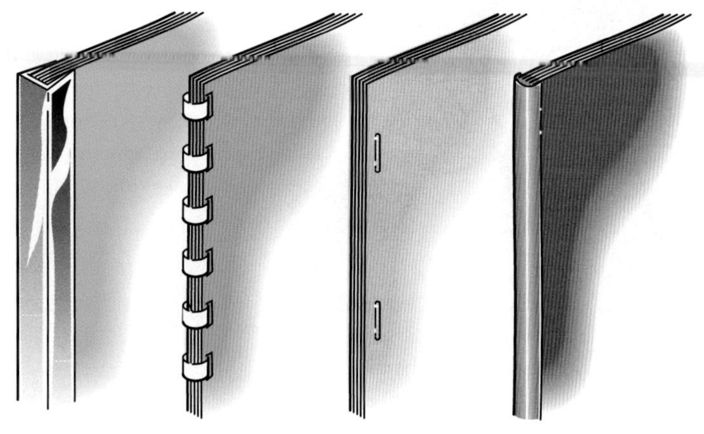

Binding methods

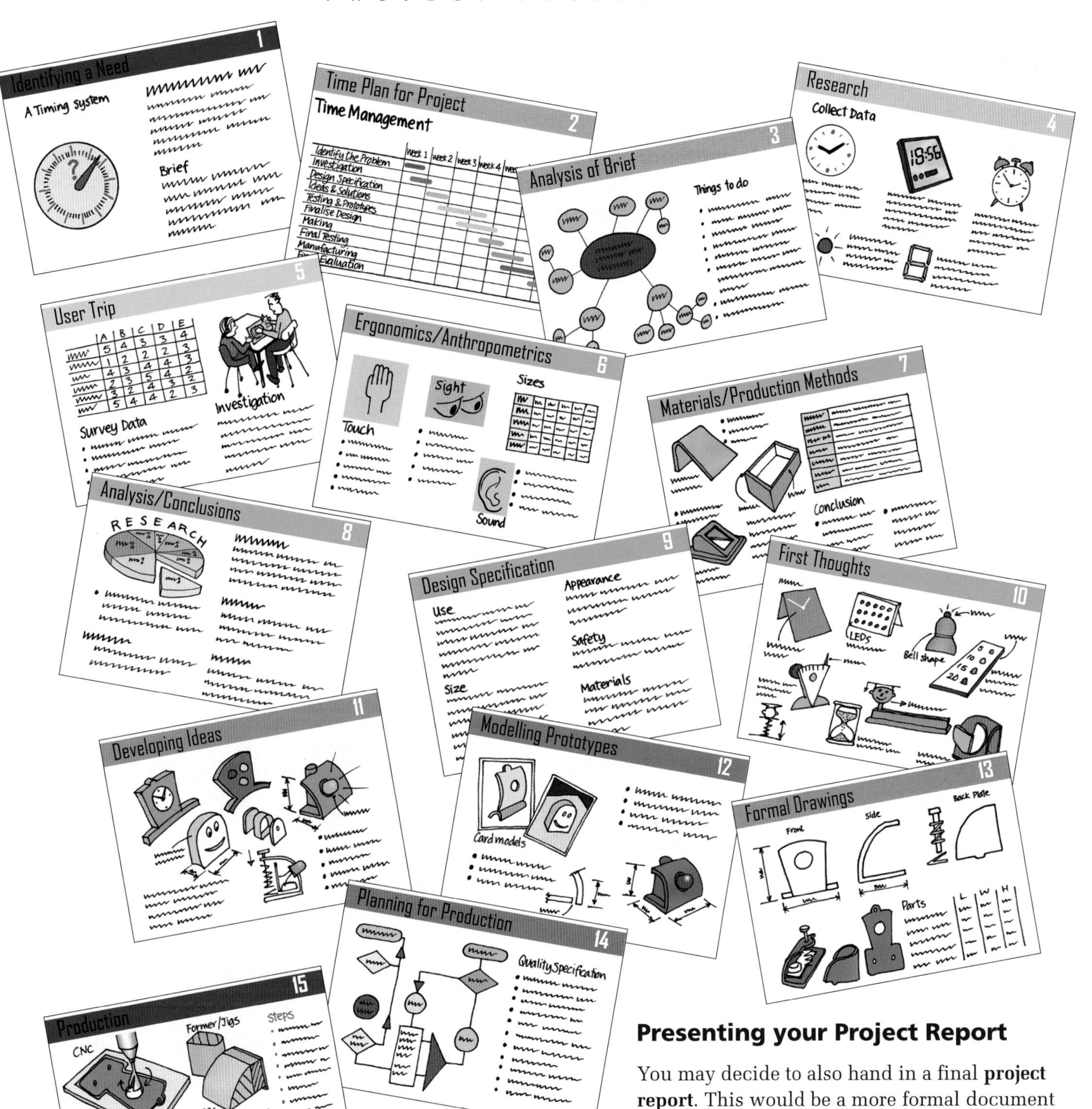

Presenting your Project Report

You may decide to also hand in a final **project report**. This would be a more formal document and used to, for example:

- describe and discuss the development process
- document detailed research material
- include an extended project evaluation
- evaluate your project in detail.

If possible, type up the report, using a word processor, or a DTP program. Remember to think carefully about the design of the layout of text, and to include illustrations such as statistical graphs and charts, technical drawings and photographs, as appropriate.

Your project report could include:

- a cover
- an introduction
- your investigation and development
- test results
- your final evaluation
- an appendix.

Project One: Introduction

An international chain of hotels needs a design for a distinctive clock to place in its public spaces and bedrooms. The timepiece will need to reflect the company's image and complement the high quality of style and taste of their hotels.

Can you design and make a suitable clock?

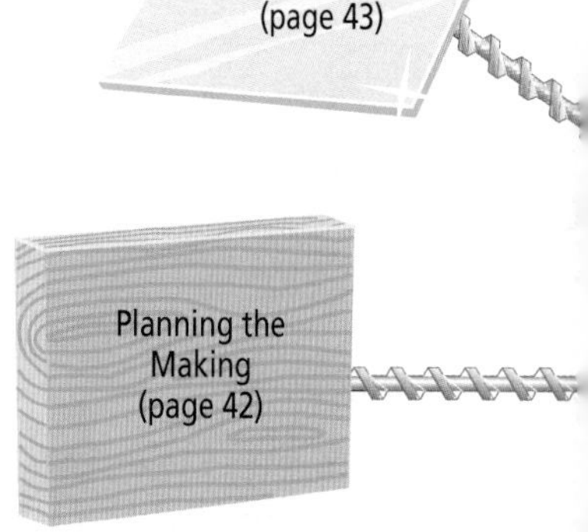

People generally stay in hotels on holiday or on a business trip. Either way they will need to be reminded of the time; perhaps they need to catch a train or flight, get to a business meeting on time, or maybe join a coach for a guided tour.

The Brief

I am writing to commission you to design a clock for use throughout our chain of 20 Future Hotels, which are situated in city centres and airports. The clock will be used in several different locations within each hotel and will need to be easy to read.

I enclose some photographs which illustrate the type of modern, efficient yet comfortable image which we aim to present to our hotel guests. Obviously the clock will also need to complement the existing interior design of our hotels.

I would like to meet you to discuss the details of the specification and look forward to hearing from you soon.

Yours faithfully,

Kevin Emit

Kevin Emit
Managing Director, Future Hotels plc

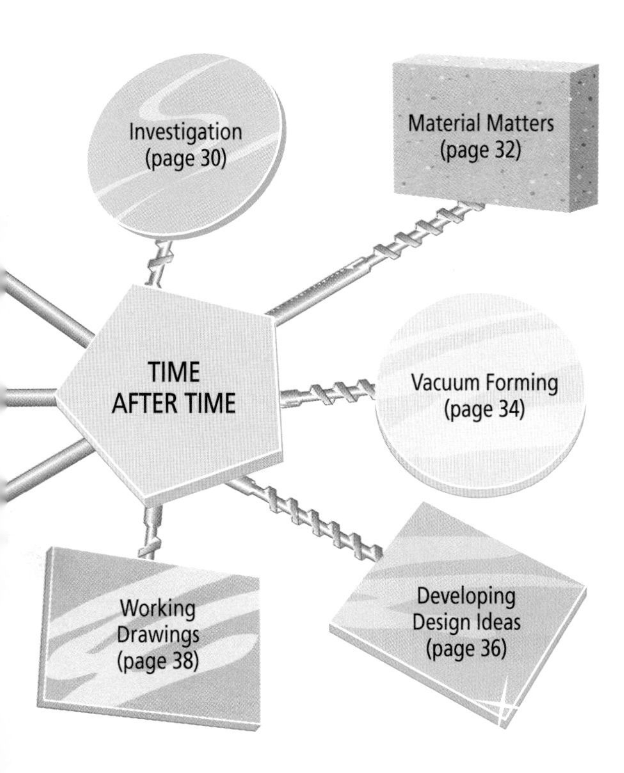

Getting Started

Read the letter carefully. What exactly have you been asked to design?

Try out some initial ideas on paper.

There are some further details that you will need to find out from the client.

NOTES FROM MEETING WITH MR EMIT:

Q: Would the clocks be hung on walls or placed on surfaces such as tables?

A: Both.

Q: Will they all be the same size?

A: No, two different sizes, one for the public spaces such as reception and the hotel bar and another, smaller size for the bedrooms.

Q. Would they be battery powered or use mains electricity?

A: Battery.

Q: Would you prefer Roman or Arabic numerals?

A: Either, or another way of marking the hours could be used.

Q. How many do you want?

A: 1000, but only one sample of the small or large clock at first, to decide whether to order the rest!

Investigation

Before you start to design your clock, you will need to ask some questions about what the clock will have to do, what it will look like and how it might be made.

In order to find the answers you should do some research.

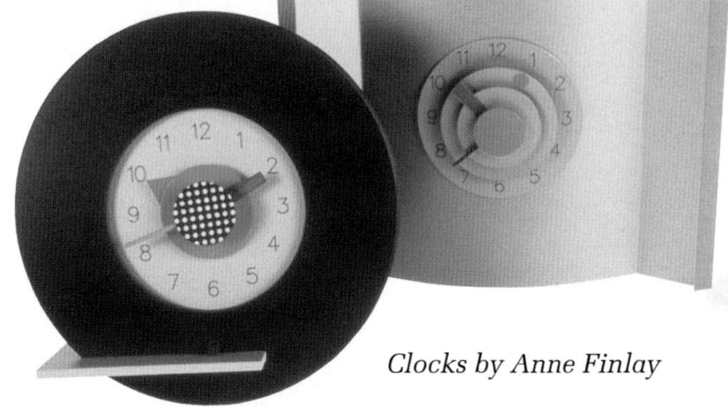

Clocks by Anne Finlay

Research Methods

You may need to gather information about clocks before you start to think about writing the design specification. You could:

- find out what eventual users think
- find out about similar products
- use a questionnaire to carry out market research
- find out about the history of your product, and how it relates to other cultures.

Think about the best way to record the results of your research: sketches and photographs, a chart or matrix, an annotated drawing, short notes taken from books, a transcript of a taped interview, etc.

Product Analysis and Evaluation

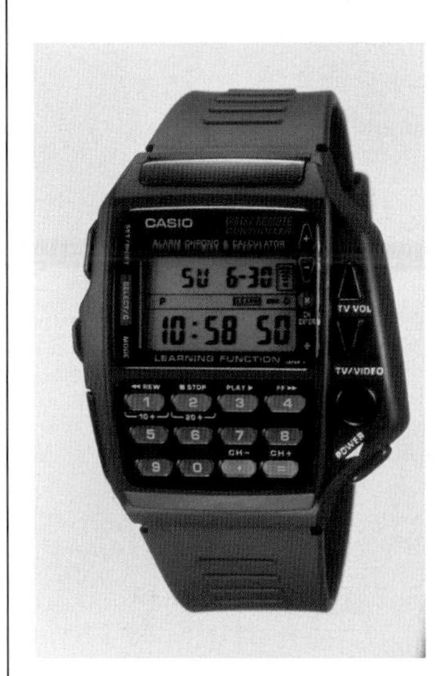

Look carefully at a digital watch and write down a series of questions which could be used to analyse all watches. Put them under two main headings:

- **Form** – that is the appearance of the watch, e.g. what it is made of, the impact of the materials on the environment, how easy it is to use, any weakness in manufacture, etc.
- **Function** – what it does and how well it does it, e.g. how durable and reliable it is, safety, etc.

Try the questions out on a selection of watches.

How might the observations you have made be useful to a designer who has been asked to design a new watch or clock?

Undertake a similar activity in order to gather information to help you to design your clock.

For information about watches go to: **www.swatch.com**

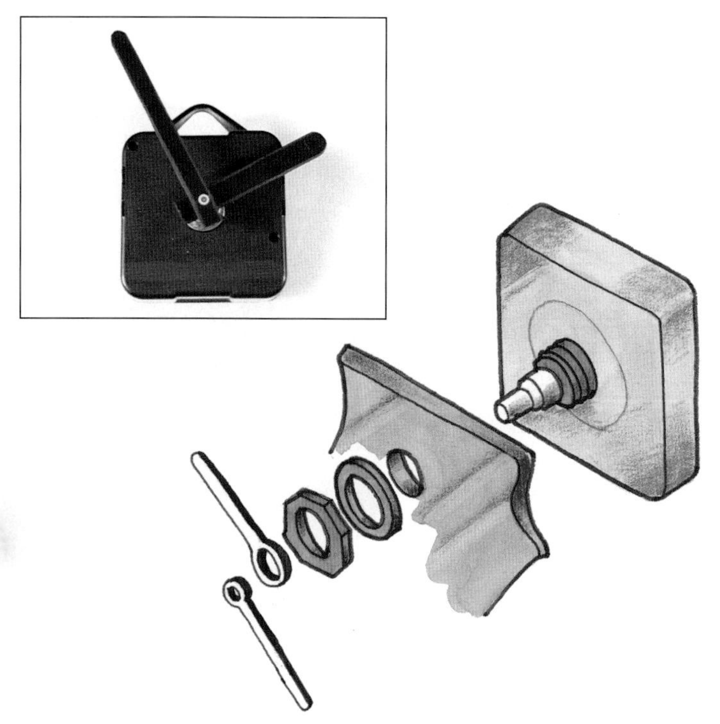

Disassembly

In your Future Hotel clock you will need to use a battery-powered clock mechanism. You will need to make a careful study of their size, shape and function, and how they can be used.

Look carefully at the battery-powered clock mechanism. Take it apart one component at a time and lay the pieces on a sheet of paper in the order in which you take the clock mechanism apart.

Sketch the components in the form of an **exploded diagram** with notes. What is the purpose or function of each component?

The distance between the hands and the casing for the clock mechanism is a very important dimension. Why do you think this is?

Another key dimension is the diameter of the protective sleeve which fits over the shaft to which the hands are attached. How will the design of this influence your design?

A short history of time

Once upon a time...ordinary people didn't need clocks and watches, the routine of their lives depended on the seasons of the year and the hours of daylight. In Britain in the 1700s there were several different time zones, but few people took much notice of these, they simply weren't affected by them.

The development of the railway system provided an efficient way of moving people and goods around the country. Suddenly it became important for ordinary people to know what time it was. As Britain became an industrialised nation, so people stopped living by the rhythms of the natural world and started to live by the clock.

At first the town hall or church clock would tell you the time, often by chiming on the hour, but gradually people needed to have their own way of telling the time, one that they could carry around with them. These devices have developed over the years from expensive gold and silver pocket watches, which men kept in their waistcoat pockets, to wrist-watches in cheap and colourful plastics.

ICT

Use the Internet and CD-ROMs to help find out about the history of time.

When the bell rings

- We are still ruled by time. Count up how many different clocks and ways of telling the time there are in your house. Don't forget the timer on the cooker, the video and maybe inside the car.
- What else can you find out about interesting and unusual **time-pieces** of the past?

Material Matters

You need to find out about the characteristics of materials which could be used to make your design, including how they may be shaped and formed. One material which you may want to use for its bright colours and durability is plastic.

What characteristics will the material that you choose to make your design from need to have?

The Characteristics of Plastics

There are seven important characteristics of **plastics**:

- **strength** (how easily it breaks)
- **toughness** (how much impact the material withstands before breaking)
- **hardness** (scratch resistance)
- **density** (how heavy in comparison to size – low density plastics float, high density plastics sink)
- stiffness and bending (how it reacts to **tension** and **compression**)
- **thermal conductivity** (how well it conducts heat – high conductivity means they are poor insulators, low conductivity means they are good insulators)
- **environmental resistance** (how easily it corrodes).

Plastics can be made clear or opaque in any colour. They are inexpensive and light, and easy to form into complex and unusual shapes and textured surfaces.

There are many different plastics. They are often called by their initials:

- LDPE, low-density polyethylene
- HDPE, high-density polyethylene
- PP, polypropylene
- PS, polystyrene
- PET, polyethylene terephthalate (polyester)
- PVC, polyvinyl chloride

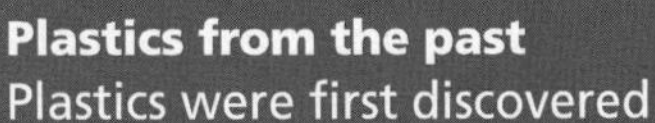

Plastics from the past
Plastics were first discovered about 1850. They were initially used to imitate other materials. Bakelite was developed during the early 1900s, and was used for radio casings in the 1920s. Perspex (an acrylic) was developed in 1936.

Today there is an enormous variety of different plastics used in almost every type of product.

Thermoplastics and Thermosets

There are two types of plastic:

▷ thermoplastics
▷ thermosetting plastics.

Both types are products of the petrochemical industry, although there are also naturally occurring plastics such as rubber. Oil, air, water and salt are processed to produce plastics in the form of powders, resins and granules.

Thermoplastics

Thermoplastics are less resistant to heat and fire than thermosetting plastics. One of their most important characteristics is that they are reshaped by heat. Unlike thermosetting plastics they do not undergo a chemical change during heating. Thermoplastic sheet can be re-heated after forming and will return to its original flat form. This is referred to as **plastic memory**.

Thermoplastic	Properties	Uses
Acrylics	Good impact strength and transparency	Goggles, lenses, windscreens, etc.
Polyamides (nylon)	Tough, resists abrasion, self-lubricating	Gears, fasteners, fishing lines, ropes, etc.
Polycarbonates	Strength and toughness. Resist impact	Safety helmets and bottles
PVC (polyvinyl chloride)	Inexpensive, can be rigid or flexible. Low strength and heat resistance	Blister packaging Outdoor use: signs, hose-pipes, cable insulation, floor tiles etc.
Polyethylene	Good electrical and chemical resistance. Can be made in different densities	Litter bins, toys, machine components, etc.
ABS	Forms well to a high definition	Dingy hulls, car dashboards, telephone handsets
Polystyrene	Easy to use. A wide range of colours available	Light fittings, vending cups, packaging trays.

Thermosetting plastics

These are stronger, harder and stiffer than thermo-plastics. They are resistant to heat and fire and are often used in situations where this is an important feature. A chemical reaction is involved in the moulding and setting of thermosetting plastics.

Thermoset	Properties	Uses
Polyester resin	strong when reinforced with glass fibre	GRP mouldings for boat hulls, car bodies.
Melanine formaldehyde	strong, hard and heat resistant	kitchen work surfaces, plastic tableware.
Phenol formaldehyde	hard and brittle	saucepan handles.
Urea formaldehyde	tough, attractive	electrical fittings.

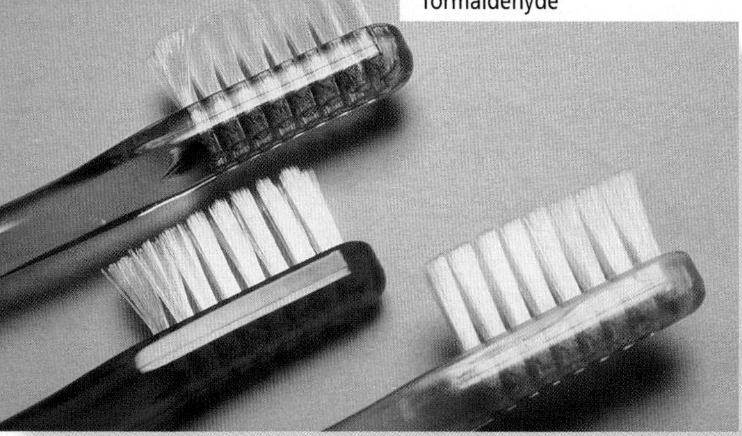

■ **ACTIVITY**

Take a good look around the room in which you are now sitting and identify at least four different items made from plastic. For each item, make a statement about why that particular type of plastic has been used for that purpose.
See if you can find out more about:

▷ the history of the development of plastic products
▷ the chemical structures of different plastics.

To find out more about LEGO go to **www.LEGO.com**
To find out more about plastics go to: **www.materials-database.org.uk**

KEY POINTS

- Plastic can be clear or opaque in any colour.
- There are two types of plastic: thermosetting and thermoplastics.

Vacuum Forming

Vacuum forming is a manufacturing process used for producing many identical copies of the same shape and form. To ensure a high quality product, some practice is needed.

Vacuum Forming

This method of forming plastic is commonly found in industry in the production of packaging, display stands, disposable cups and trays and other relatively cheap items. It is also a process which is commonly used in schools. This means that you can use a similar process to one used in mass production in industry.

Suitable materials for vacuum forming

There are several thermoplastics which are suitable for vacuum forming. These include:

- polystyrene
- ABS (acrylonitrile butadiene styrene)
- PVC (polyvinyl chloride)
- polypropylene and acrylic sheet (one trade name of which is Perspex).

They all come in a range of colours and gauges (or thicknesses).

The vacuum forming process

This is a simple process in which air is sucked out from underneath a plastic sheet which has been softened by heat. This pulls it down onto a mould which is the shape of the product which you want to make. The excess material is then trimmed away and the edge finished.

Using the vacuum former

1 The mould for the shape that you want to make has to have slightly sloping sides. If it has a large base the mould should also have very small holes drilled in it. French chalk can be used to provide a releasing agent.

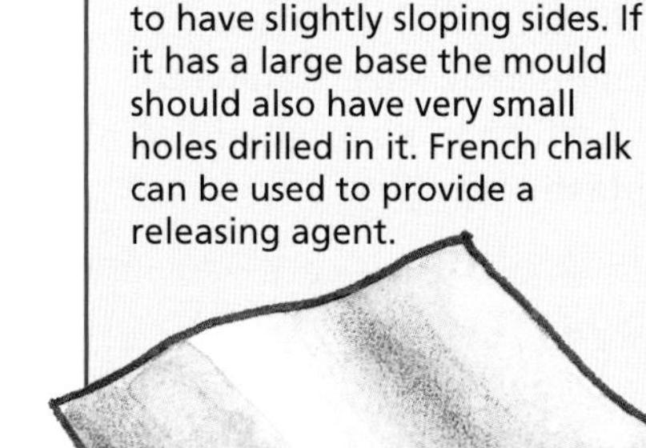

2 The mould is placed on the platen inside the vacuum former which is then lowered.

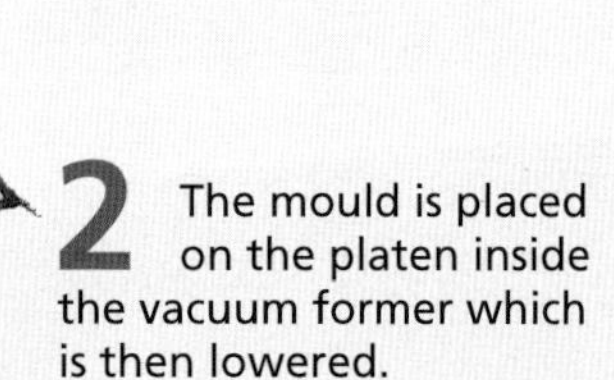

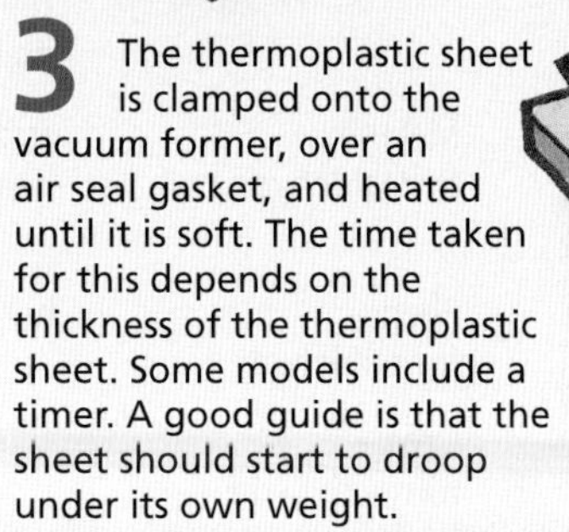

3 The thermoplastic sheet is clamped onto the vacuum former, over an air seal gasket, and heated until it is soft. The time taken for this depends on the thickness of the thermoplastic sheet. Some models include a timer. A good guide is that the sheet should start to droop under its own weight.

4 The platen is raised, bringing the mould up to the softened thermoplastic sheet.

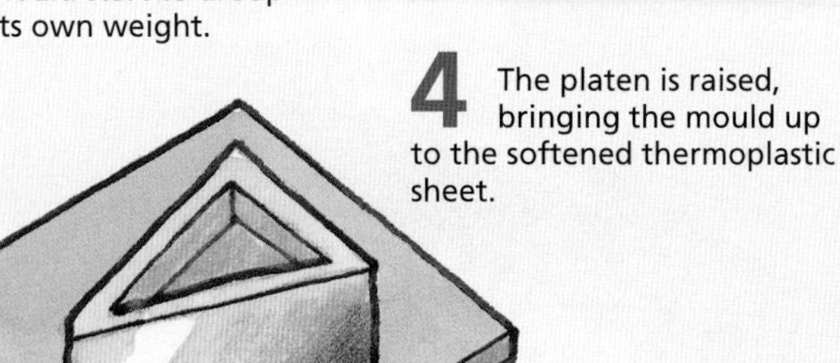

Trimming and finishing

Cut away waste material using a scalpel, a safety ruler and a cutting mat. Your fingers should form a bridge over the cutting line. The blade should be at an angle to the plastic.

The final part of the process is to remove the waste material. This is carried out using a scalpel or Stanley knife, a bandsaw, shaper saw or scissors.

The edge is then smoothed with wet and dry paper.

Waste material can also be removed using a horizontal slitting saw such as the Gerbil. It produces a flat, well finished surface.

Care should be taken as sometimes the cutting action is too aggressive and results in a chipped edge. Try with a piece of waste thermoplastic material first, and if this happens, reverse the rotation of the saw to a less severe rubbing action.

5 The air underneath it is removed by vacuum pump, pulling the sheet down over the mould.

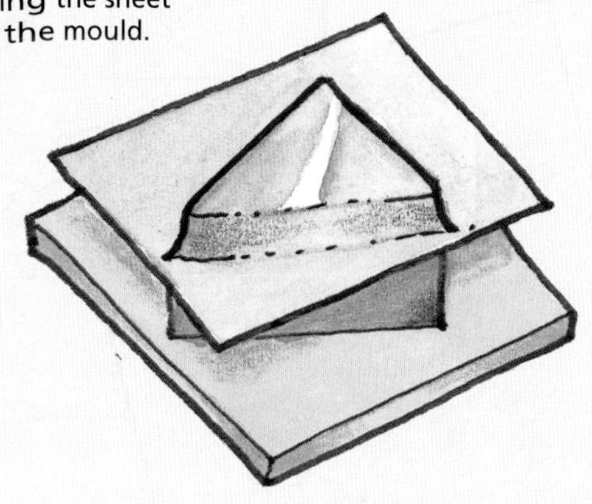

6 The vacuum pump is then used in reverse to blow air in order to separate the plastic sheet from the mould. The work is allowed to cool and removed from the machine.

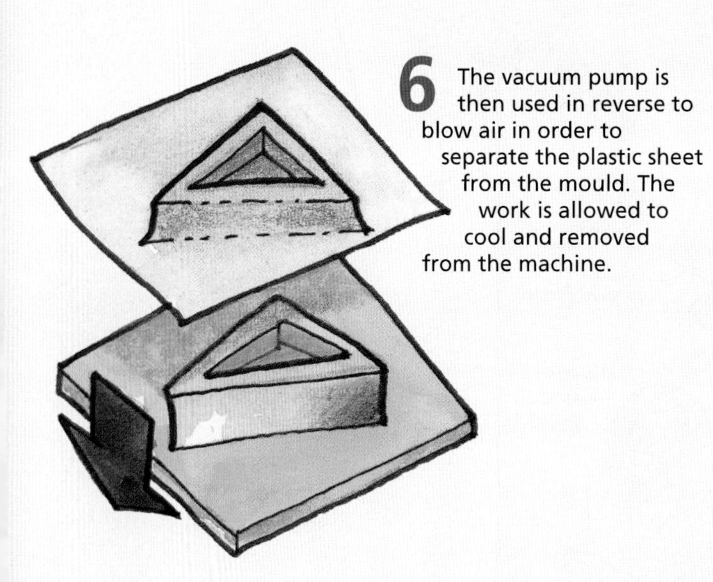

If it doesn't work...

- Make sure that there is an airtight seal all the way around your material before you start to vacuum form it. If necessary adjust the clamps which hold the frame onto the gasket.
- Make sure that the thermoplastic is sufficiently hot and therefore pliable, before switching on the vacuum pump.
- Make sure that you raise the platen before switching on the vacuum pump.
- Make sure that if the mould has a thick and/or deep base, you have drilled some very small diameter holes in it to help the thermoplastic to mould over it.

You may not get the process right the first time. Try to work out what went wrong and make any necessary changes before trying again.

Common problems are:

- the thermoplastic wasn't hot enough before the platen was raised and the vacuum pump switched on.
- the mould needed additional air holes to allow the thermoplastic to be pulled down more tightly.
- the thermoplastic wasn't clamped tightly to the machine.
- the sides of the mould didn't slope enough to allow for easy removal.

Developing Design Ideas

Once you have written the* design specification *and researched the materials and processes which you might consider using, you can make a start on your design ideas in more detail. Don't limit your ideas at first. Be as adventurous and experimental as possible.

Design Specification

- The clock will be placed...
- The smallest size it can be is...
- It should not be larger than...
- It will be made from...
- Its colours and textures should be...
- The hours will be marked by...

Getting Started

You might try :

- picking out the key words from your specification e.g. 'modern', 'futuristic', 'efficient', 'easy to maintain'.
- making a spider diagram which links these words and may lead to the generation of more ideas.
- imagining what it would be like to be in the hotel. What sort of clock would you expect to see there?
- looking at the clocks which you sketched right at the beginning of the task. Could they be adapted in some way?
- modelling in card or other cheap materials to experiment with some 2D shapes for your clock and 3D form. Investigate how it should stand up, how it will accommodate the mechanism and how far away from the wall it will project if it is to be hung on a wall.

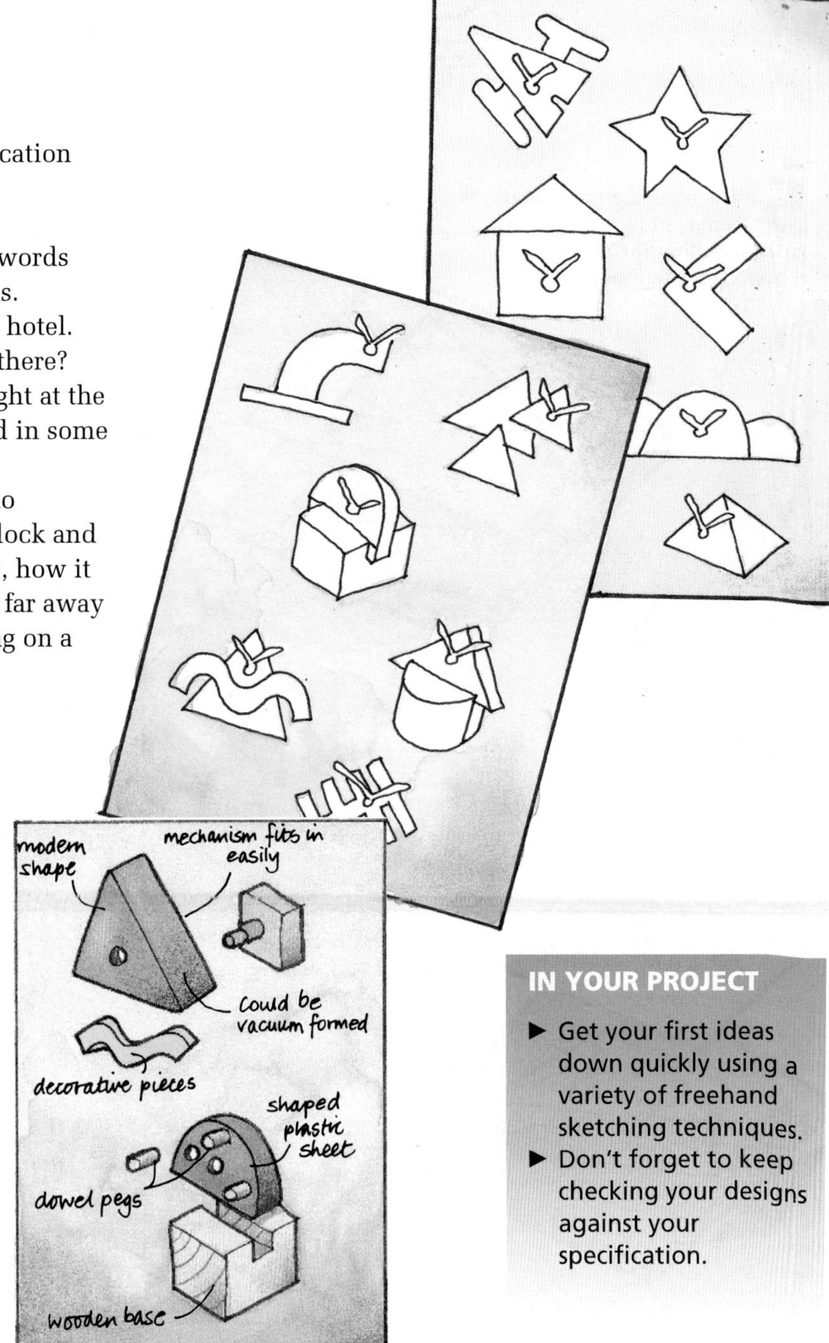

Second Thoughts

Look again at your specification. Which of your ideas reflects the statements which you wrote?

Choose two or three of these ideas and develop them further in freehand sketches.

Annotate your designs with notes about possible materials and components that you might use and how you might make your design.

IN YOUR PROJECT

- Get your first ideas down quickly using a variety of freehand sketching techniques.
- Don't forget to keep checking your designs against your specification.

Planning and Making Prototypes

A **prototype** is a form of mock-up or model of your design. It can be used to make initial tests to guide your designing. A prototype is usually very similar to the idea you have drawn, but helps highlight problems which are not apparent from a sketch.

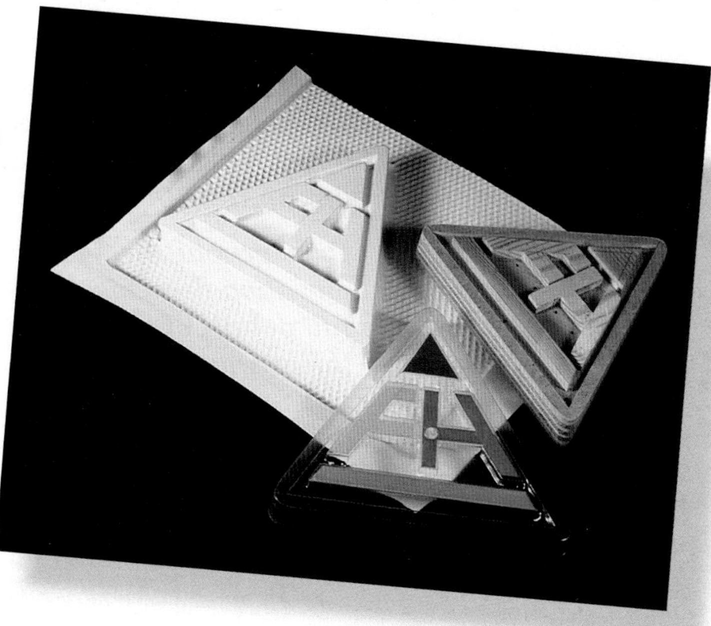

What sort of things might you want to test about your clock design? Remember that you need to think about the way in which the clock will be used as well as what it looks like.

- ▷ How easy is it to tell the time?
- ▷ How well does your clock perform at 1 metre, 2 metres, 5 metres?

You might want to consider:

- ▷ the size of the hands
- ▷ the size of the hour markers or numbers
- ▷ the proportion of the clock face relative to the rest of the clock.

What else might you be able to test?

Don't forget to check your prototype against your specification.

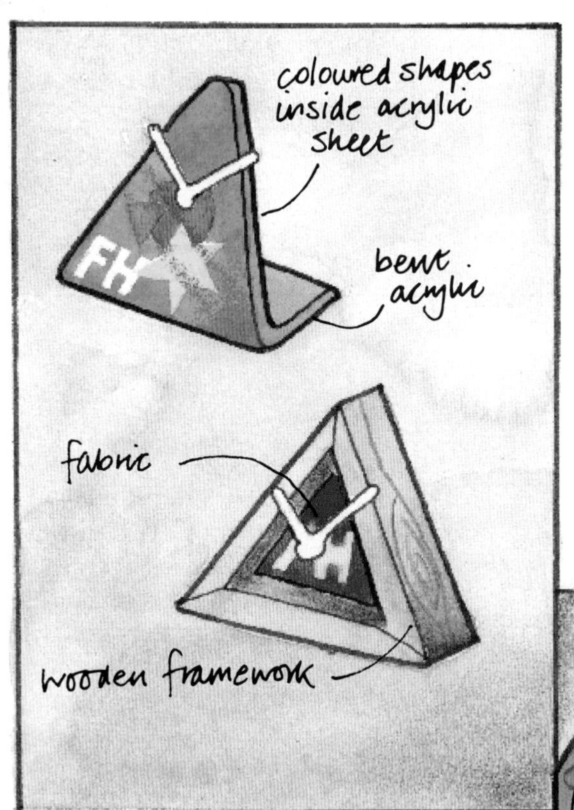

Extending Your Choice of Materials

Before you finally decide which materials you are going to use you will need to check on their availability with your teacher.

Don't forget to think about different effects which you might achieve by combining unexpected materials, for example clear acrylic and cellophane or wood with textiles. You will have to research techniques for joining different materials and include notes about this on your design drawings.

Don't always decide to use adhesives. Sometimes mechanical fixings such as nuts and bolts or screws will perform better and can become an effective visual feature of the overall design.

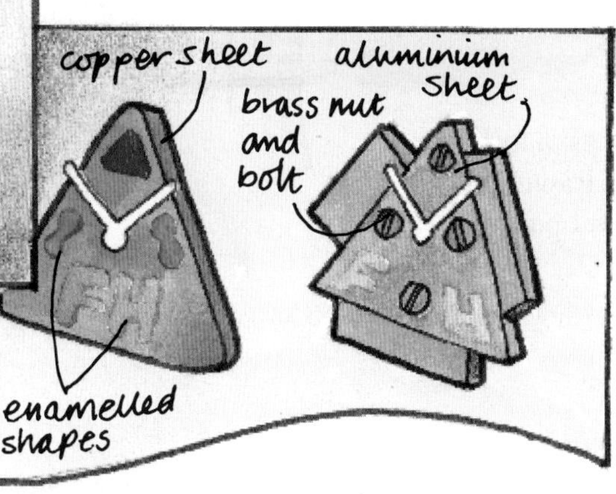

Working Drawings

When you have chosen your final idea, working drawings need to be prepared. Think about which technique would be the most appropriate for your design.

Orthographic drawings combine plan and elevation drawings of an object. They provide full details of its size in three dimensions and the arrangement of the various parts.

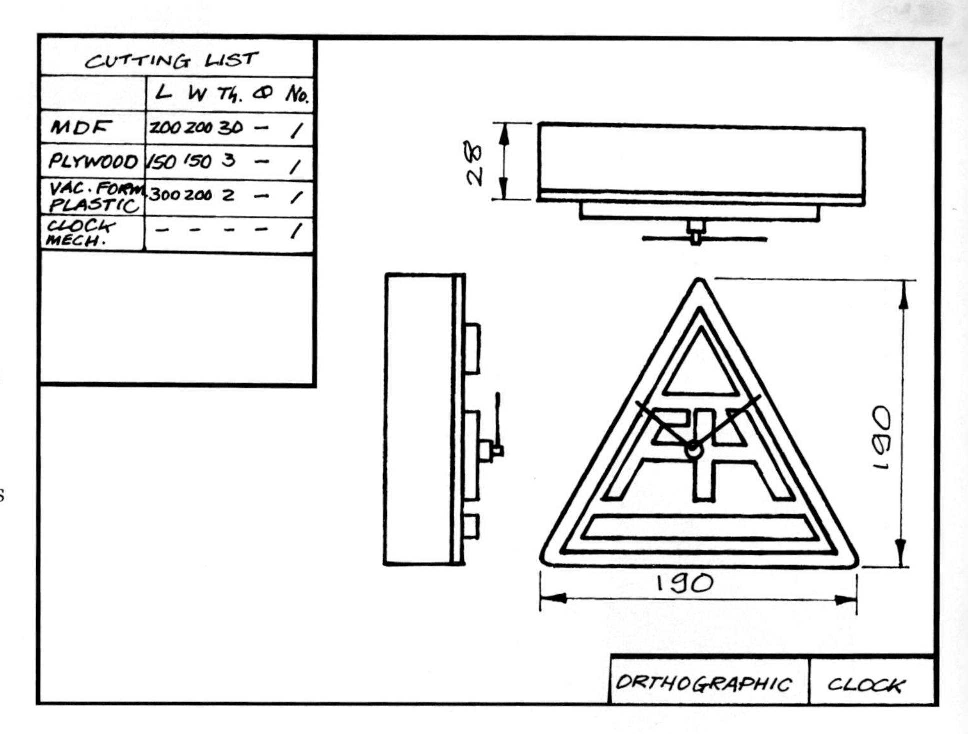

Orthographic Drawings

The arrangement of **orthographic drawings** and the method of dimensioning them are defined by a British Standard. The standard provides a common method for everyone to follow.

The drawings should always be placed so that the plan and front and side elevations line up exactly. The number of views shown should be the minimum needed – usually three.

In a **general arrangement drawing** the product is drawn in its final assembled form. Overall dimensions are given, and each component is numbered.

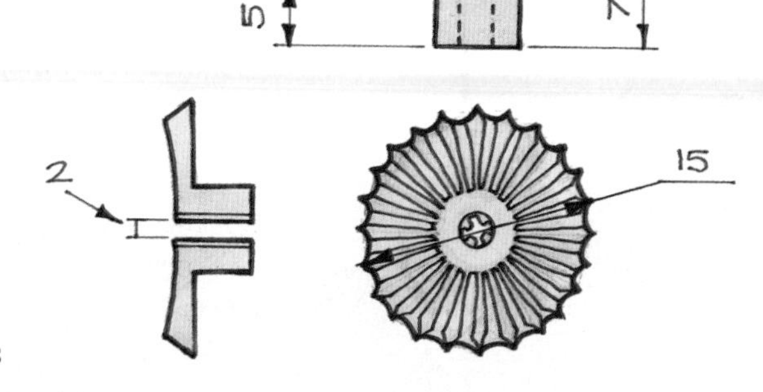

Detail drawings

On a series of detail drawings, a fully dimensioned orthographic drawing is provided for each component part.

■ ACTIVITY

Go back to the drawings of the range of clocks which you sketched at the start of the project (see page 29) and make a general arrangement drawing of one of them. Exchange your drawing with a friend and see if they can make an accurate copy of the drawing without you providing any verbal instructions.

Dimensioning

There are some clear rules for adding sizes to drawings:

- ▷ All dimensions are given in millimetres, and are written as the number only.
- ▷ Numbers should be placed directly above the middle of the dimension line.
- ▷ Numbers should be placed so that they can be read from the bottom or from the right of the drawing only.
- ▷ The minimum number of dimensions should be included.
- ▷ Dimension and projection lines should be half the thickness of the lines of the object.
- ▷ There should be a small gap between the end of the projection lines and the object.
- ▷ Arrows should be placed accurately between the location lines, and be filled in.

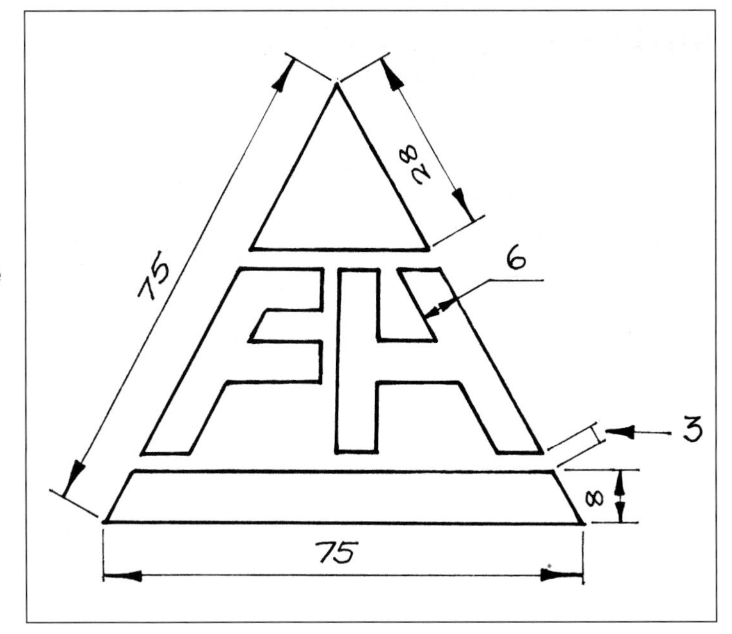

Consult BS308 or PD7308 for fuller details of the system which should be followed.

Computer-aided Drawing

There is a wide range of **CAD** packages which enable complex and accurate orthographic and detailed drawings to be prepared. These are now used extensively in industry to save time and money.

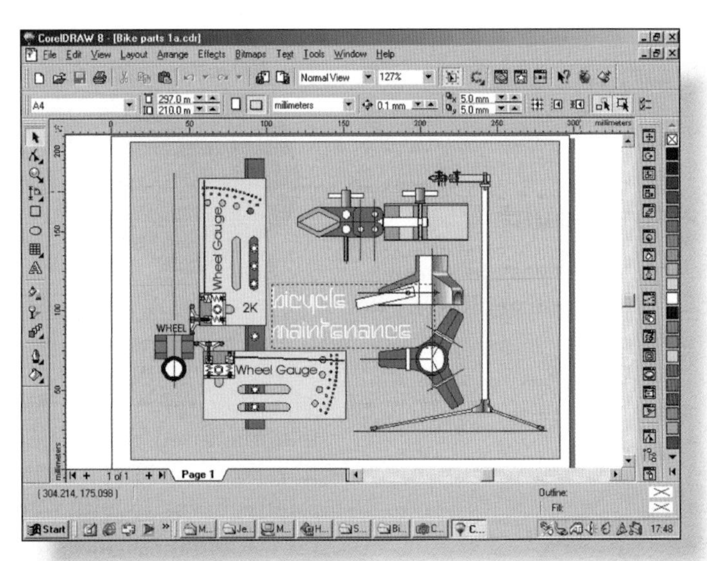

IN YOUR PROJECT

- ▶ You should use an orthographic projection for your final working drawing.
- ▶ The layout and **dimensioning** of orthographic drawings must follow BS308 or PD7308

These parts for a bicycle wheel gauge were designed using CAD and made from the drawings using CNC machines. The 'wheel' parts are aluminium and the gauge is 3mm PVC. When the drawings are converted to code for the CNC machines all the shading has to be removed and the parts have to be separated.

KEY POINTS

- Make sure that you use appropriate formal drawing techniques for your working and presentation drawings.
- When you are exploring your early ideas, don't use formal drawings. These aren't useful when you want to get your ideas down quickly.

Designing and Making a Mould

Before you can vacuum form your final product, you will need to design and make a mould. This is the same shape as the object which you want to make. It needs to be made very carefully to ensure that it performs well and produces a high quality product.

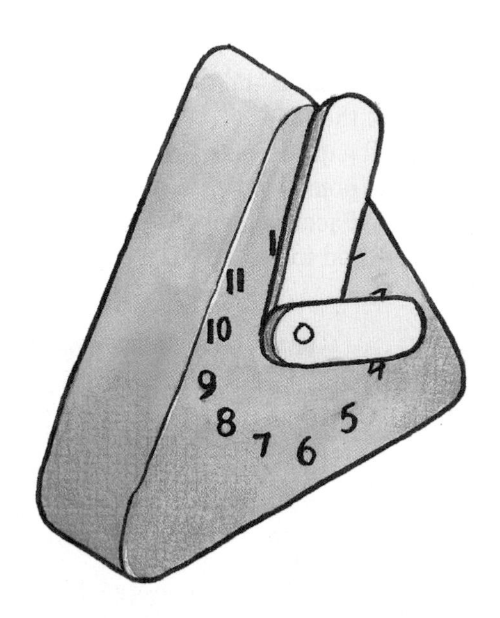

Designing the Mould

When you design your **mould** think carefully about the size and proportions of your finished design. At this stage it is easy to adjust the shape and form of your mould. Once the plastic has been formed it is very difficult to change.

Choosing Material for the Mould

You will need to use a material which is easy to shape and has either a very fine grain or no grain at all. Why do you think that this is important?

There are a number of materials which are suitable. These include woods (for example Jelutong) and manufactured boards (for example medium density fibreboard or MDF). Both of these materials are worked relatively easily with traditional woodworking and power tools. They remain stable at 160 °C and are therefore able to withstand the temperature of the vacuum former heating elements.

If blocks of material are to be joined, Cascamite or PVA glues should be used. The drying time may be reduced by heating in an oven at 60 to 80 °C.

Other suitable materials include card, metal, plaster of Paris, dental plaster, epoxy resin and clay.

Adding Texture

In order to produce a texture on your finished product, you can place perforated sheet metal or card, mesh or even glass paper between the former and the thermoplastic. The easiest way is to place it on top of the mould before lowering the platen into the machine.

On the Edge

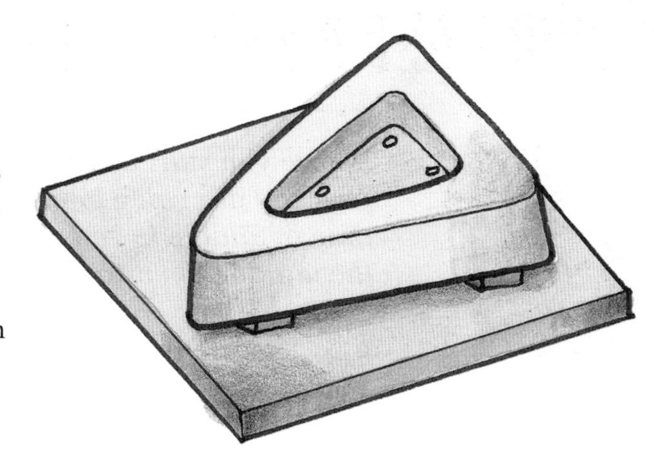

To achieve neat edges and avoid 'webbing' at the corners, place the mould on a base board cut to fit onto the platen and drilled with a hole to allow the air to be removed when the vacuum pump is switched on.

Fixing the mould to the baseboard with adhesive pads will separate the two, providing an air passage between the baseboard and the mould. This will allow more accurate trimming away of waste material.

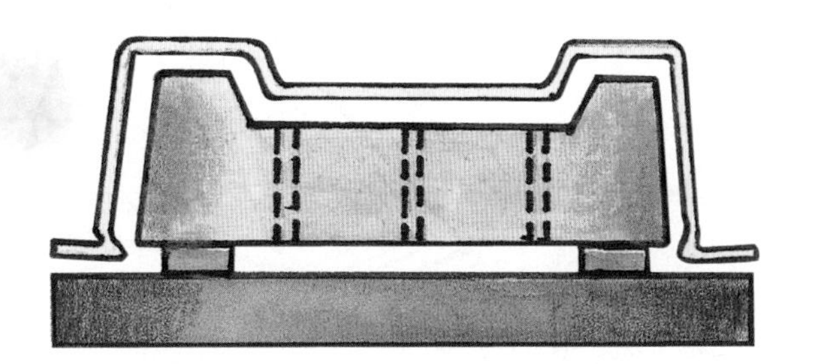

Very small diameter holes (i.e. 1.0 to 1.5 mm) should be drilled through the mould if there are cavities where the plastic sheet might form a seal when the air is pumped out. This can be done with a high speed PCB (printed circuit board) drill.

A piano wire drill may be used to complete the drilling process where the depth of the hole needed exceeds the length of the drill bit. Alternatively drill a larger hole from underneath, and drill the small hole from the top.

IN YOUR PROJECT

If your mould doesn't work, refer back to page 35 to check out some of the common problems found with vacuum-forming.

KEY POINTS

- Make sure that you drill holes to allow air to be evacuated.
- Radius edges and use releasing agents to ensure easy release of the mould from the plastic.

Ensuring an Easy Release

Make sure that any vertical sides on your mould are made to slope slightly. This will enable you to remove the mould more easily when the vacuum forming process is complete.

All corners and sharp edges should be radiused to allow the mould to be removed without difficulty. Ask about suitable release agents, such as French Chalk.

If moulds are 'forced' from thermoplastic sheet, the product often develops white 'stress' marks that spoil the appearance of the final product.

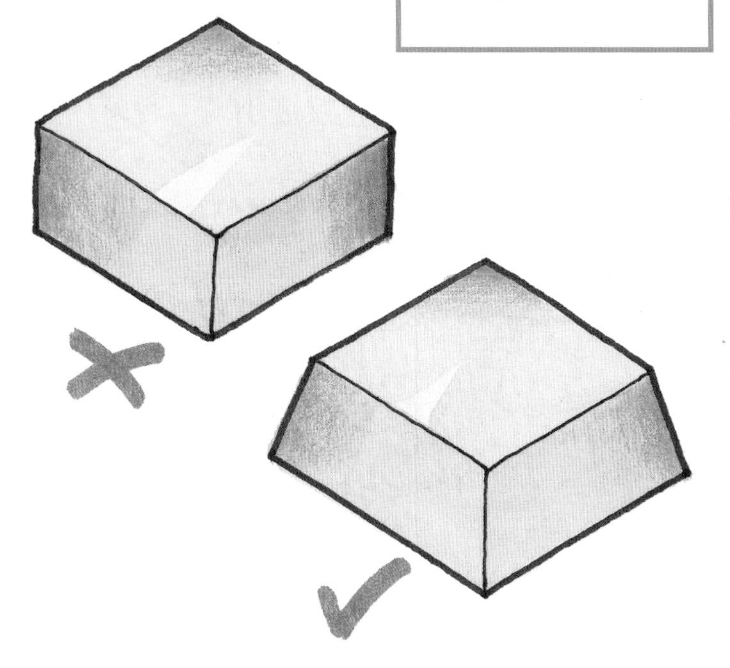

In industry

In industry a pattern maker makes moulds and formers. This is a very highly skilled job. The appearance and performance of the final product depends on the pattern maker's ability to make a mould or a former which can be used many times.

Planning the Making

Before you start making you must plan exactly how you are going to proceed. Think about the time available to you and plan each stage of the making to ensure that no time is wasted.

28
190
1. make former from MDF.
2. make numbers from self-adhesive plastic using CAM.
3. add texture to former using card offcuts.
4. vacuum form thermoplastic.
5. drill hole for axle.
6. add painted decoration, test.

Planning and Making It!

Make sure that you have a final drawing for your design. The drawing should clearly show how any decoration or detail will be applied after the main manufacturing processes have been completed.

Prepare a list of all of the different materials, processes and making stages which you will need. Draw a **flow chart** to illustrate what you will do and when.

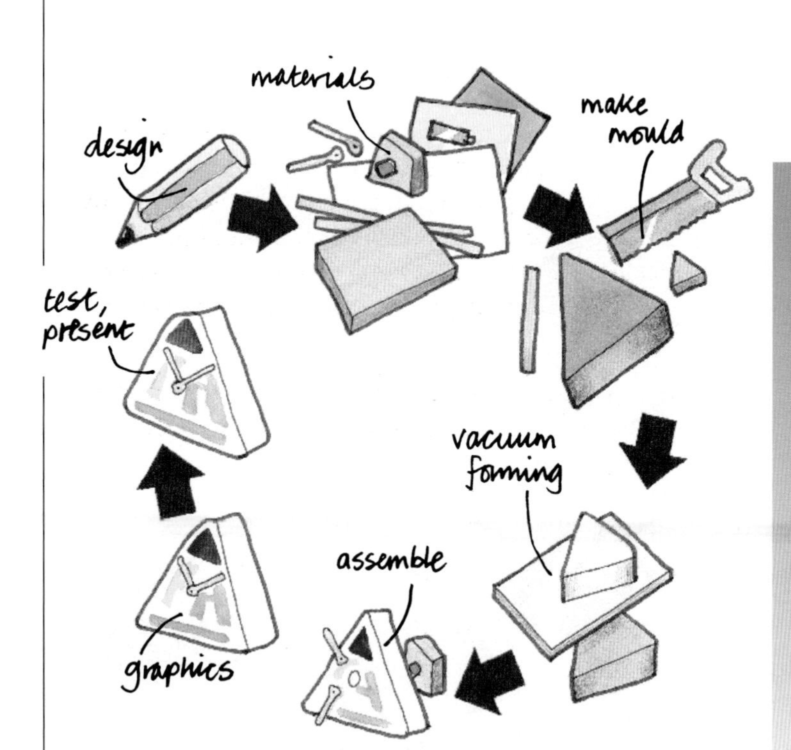

IN YOUR PROJECT

- Allocate the time that you have remaining for this project. Try to be realistic about how long each stage will take. Don't forget that you will need to allow time to test your design and write an **evaluation**.
- Are there any making stages which could be going on at the same time to save time? For example if you need to glue several blocks together to make a former, the adhesive could be drying while you make another part of the design.
- What **tolerances** (see page 138) will you be working to and how and when will you check these? This is particularly important if you are working with more than one material and they have to fit together.
- How can you ensure a high standard of finish?
- What safety precautions will you need to observe during the making process? Highlight possible hazards.

Testing and Evaluation

When you have finished making your clock you will need to test it against the specification.

When you have tested your clock you will need to write your evaluation report. This should mention both the process of designing and making, and the results of testing the clock.

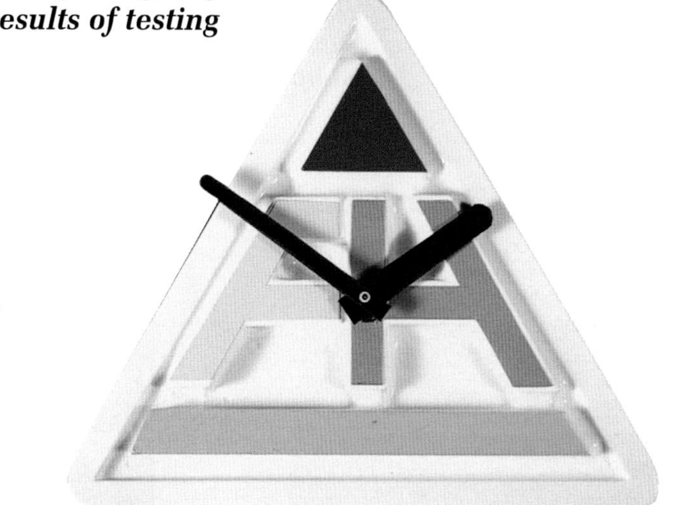

Final Testing

Checking the design

Look back at your design specification. How well does your solution meet the requirements of the specification?

- Is it the right size for its final location?
- Does it hang safely on the wall or have stability on a flat surface?
- Can the time be read easily from a distance appropriate to its location?
- Do the shape, colours and textures complement and enhance the Future Hotel chain's present image?
- Does the mechanism fit correctly?
- Will it be easy to clean and maintain?
- Will the materials which you have used result in a durable clock which will survive the rigours of hotel use?

You may have other requirements of your own to add from your original design specification.

Checking the making

You should consider the quality of finish you achieved with your making. This will involve examining how well you have done things such as:

- removing excess or waste material, for example when vacuum forming
- smoothing rough edges, for example making the edge of a piece of sheet acrylic as shiny as the other surfaces
- applying adhesive neatly to join materials, for example wood and vacuum forming sheet
- applying any protective or special finishes, such as stain, paint or varnish, to the surface of your design.

The Evaluation Report

When you write your evaluation report remember to mention good and bad points about the processes that you used. Include how you overcame problems and any major changes which you made to your planning. You may be able to suggest improvements in the processes which you used, which you would follow were you to make the clock again.

Also describe how well the clock works, including the results of the tests which you carried out. How could the clock be improved?

Dear Mr Emit,

Please find enclosed the plans for the proposed clock for Future Hotels. As you will see the design is highly distinctive and in keeping with the image of the company.

I also enclose costings to show how much it would cost to produce the number you require.

I very much look forward to hearing from you.

Yours sincerely

Project Two: Introduction

Counting the Cost (page 68)

Finishing Off (page 64)

Joining and Fixing Materials (page 62)

You have recently set up your own small business to design and manufacture jewellery on a small scale for retail outlets. The buyer for a chain of gift shops based in a variety of wildlife parks sees your work at a craft fair. He invites you to design and make a range of jewellery with a theme related to wildlife or endangered species, to be sold at their park shops.

Tuesday 19th

Finalise accounts. Order new stock.

Wednesday 20th

Meeting with Mr Fenton, Wildlife World to discuss the possibility of designing a range of jewellery for their gift shops. 10.00 a.m. (Take samples of previous work.)

Thursday 21st

Studio day

Clarifying the Brief

- What exactly has Mr Fenton asked you to design and make?
- What design features must it have?
- What features are up to you to suggest?

'Obviously It's important that the jewellery must have a wildlife theme. We thought it might reflect the culture of the country of origin of the animals e.g. tiger – India, zebra – Africa, etc. We imagine that colour, texture and pattern will be strong elements in the design.

It is essential that the jewellery appeals to male and female customers and sells for a low enough price to attract 'impulse' buyers. Initially we would need 100 pieces, so your final costings will need to be based on a method of small batch production.

We are not necessarily looking for rings and necklaces. You are welcome to suggest other types of body adornment for us to consider, or to perhaps consider at a later stage.

Whatever you come up with, the final product must be presented on or with a small card which includes either the wildlife park's name or your company logo.

To begin with we will need to see a one-off sample of your design proposal to enable us to decide if we want to go ahead.'

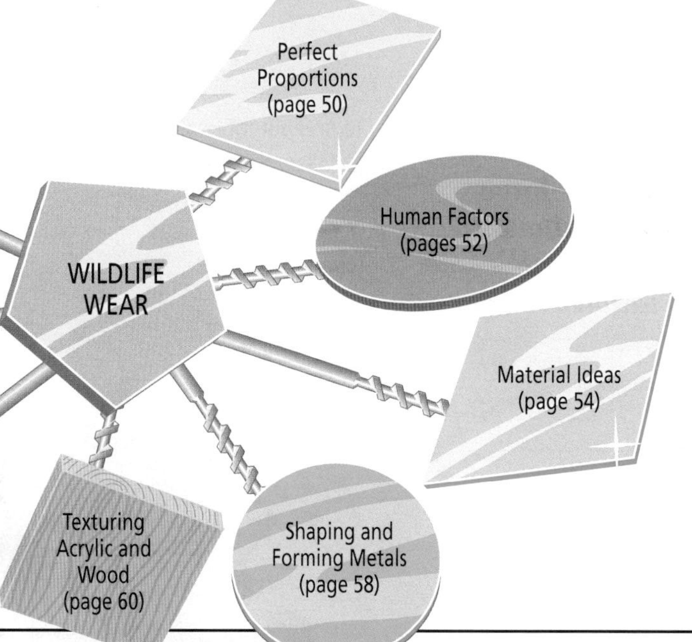

First Thoughts

- What wildlife themes could be developed?
- Where can you acquire information about wildlife parks?
- Which materials could you use?
- How will the jewellery be made?
- Who will buy or wear the jewellery?
- What price will the jewellery sell for?
- How will the jewellery be presented for sale?

Make sketches of your first ideas. Include notes to explain your ideas.

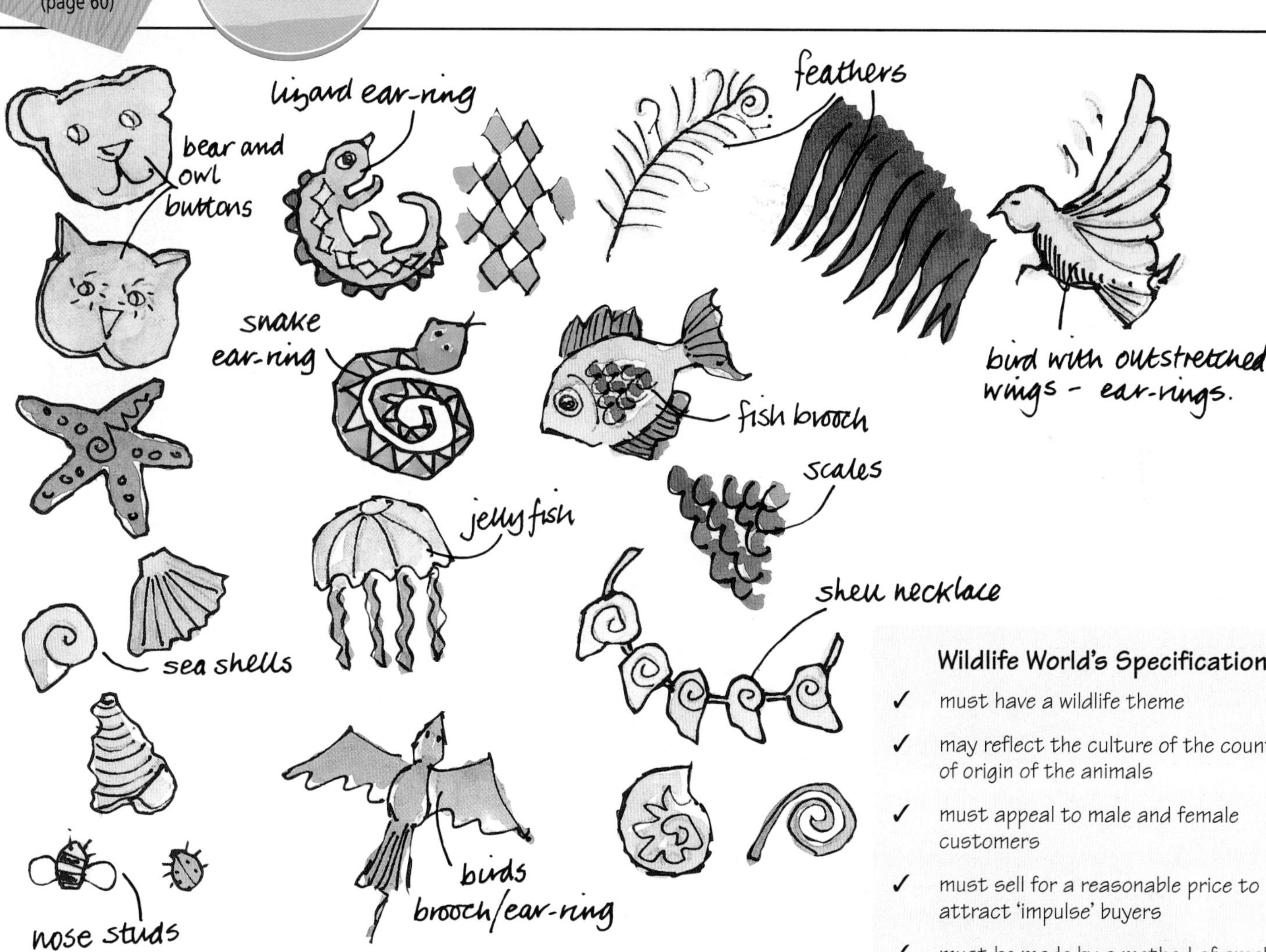

Wildlife World's Specification

- ✓ must have a wildlife theme
- ✓ may reflect the culture of the country of origin of the animals
- ✓ must appeal to male and female customers
- ✓ must sell for a reasonable price to attract 'impulse' buyers
- ✓ must be made by a method of small batch production
- ✓ must be presented on or with a small card which includes either the wildlife park's name or my logo

P.S. Phone Mr Fenton next Tuesday to arrange a meeting to discuss my first ideas.

The Body Beautiful

To find out more about jewellery through the ages go to the Victoria & Albert Museum at:
www.vam.ac.uk

For thousands of years human beings have expressed their individuality or membership of a group by adorning their bodies with decorative materials, paint or tattoos.

What is Jewellery?

Jewellery is not just objects made from precious materials, such as gold, silver, platinum and precious stones. Jewellers have always been inventive in their use of materials, particularly in the last few decades. By experimenting with an ever widening array of material contemporary jewellers produce an extraordinary selection of exciting designs.

Today antique, classical and experimental jewellery are worn at the same time. Jewellers enjoy pushing the boundaries and the challenge of developing and constructing new designs.

The most important guideline however is that the jewellery should be wearable.

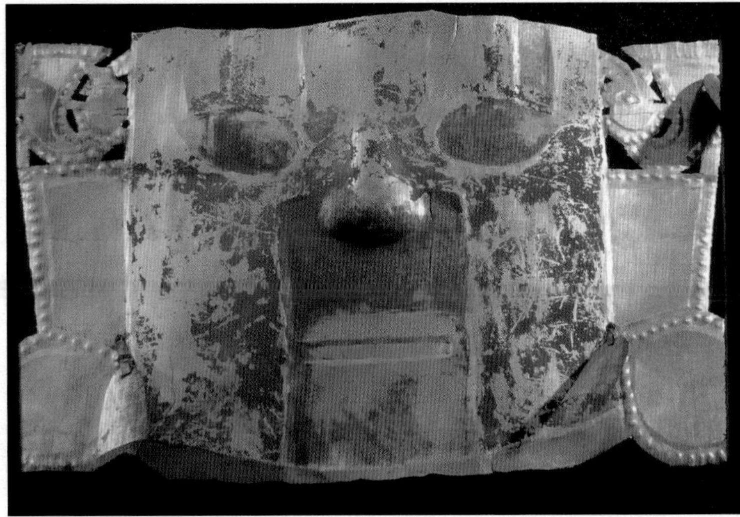

Gold funeral mask from Peru

Jewellery Through the Ages

Archaeologists studying cave dwellers of 40,000 years ago have discovered the claws and teeth of prey strung on natural fibres to form necklaces. These were worn by the hunter-gatherers of that time.

Much later European kings and nobles commissioned goldsmiths to produce beautiful crowns and tiaras fashioned out of gold, silver and precious jewels – diamonds, rubies, emeralds and pearls.

Other civilisations valued many different local materials – semi-precious stones, shells, seeds, bone, glass beads, even straw!

Pre-Colombian civilisations used colourful feathers for their head-dresses and necklaces.

North American Indians contrasted natural turquoise stones with red coral set in fine silver.

Indian jewellers have enjoyed a long reputation for producing fine jewellery made from precious stones set in gold, as well a more affordable pieces made in silver, glass beads, shells and bone. These items are often found for sale in fashion accessory shops today.

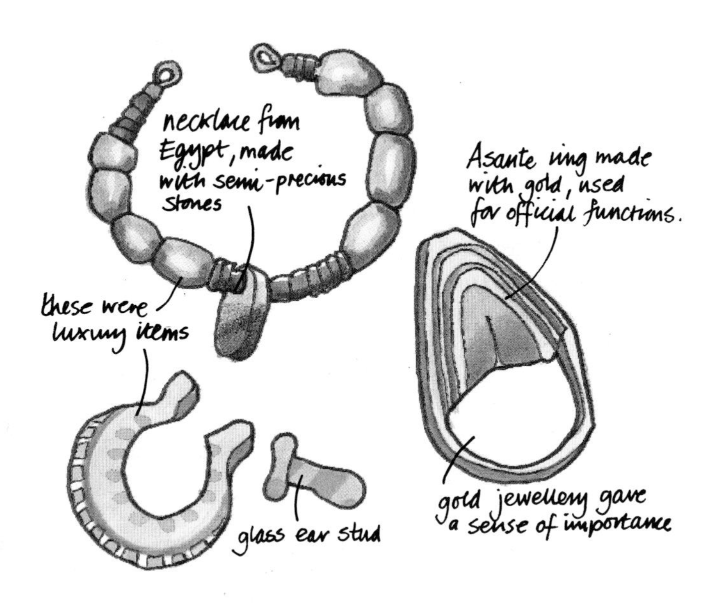

Investigation

Visit or telephone your local tourist information centre or travel agent. Collect a variety of leaflets concerning a broad range of wildlife parks.

Go to the library and find books about endangered species and the cultures of other continents, e.g. Africa, Egypt, Australia, South America, etc.

Make a study of historical and modern-day examples of body adornment. Sketch several examples and add notes which cover:

- when and where the piece was made
- the materials it is made from
- how it was made
- the sort of people it would be worn by
- the sort of occasions it might be worn on.

Gold and pearl earring from Ukraine

Emblem of the Roman Emperor Claudius

Art nouveau pendant in enamel and soapstone

Getting Started

Starting a new project can be a daunting experience, even for experienced professional designers. It is important to spend time just thinking around the subject. Jot down ideas as they come to you but always be willing to look beyond your first thoughts.

Inspiration!

Where do designers get their inspiration from? The answer is anywhere and everywhere – so keep your eyes open at all times!

- ▷ Collect magazine clippings that appeal to you – you never know when they might be just what you are looking for.
- ▷ Look out for greetings cards, birthday cards, wrapping paper, and post cards that could inspire you.
- ▷ Look at patterns on textiles and decorative articles – for example, patterned carpets such as Turkish Kelims, or Islamic architecture, fancy Victorian tiles and wrought iron work. Sketch the details that catch your eye.
- ▷ Gift shops, jewellers and fashion accessories in clothes shops are all good places to find out the latest fashions.

Keep a sketchbook of ideas and references – jot down a few notes to explain your thoughts.

Presenting Your Investigation

Assemble all the relevant information you have collected. Present your **research** clearly, explaining its importance to your project.

If you are including pictures cut from magazines and catalogues make sure you add comments which describe and evaluate the items. Point out areas of interest – historical details, repeat patterns, colours, materials, finishes, how they are made, etc.

Provide details of where the information has come from – names of books, galleries you have visited, people you have spoken to, etc.

Identifying Your Market

Conduct a **survey** to find out if males and females have different reasons for buying and wearing jewellery and how much the customer might be willing to spend. You will need this information to help you design your jewellery for your potential customers.

Present your findings in a visually interesting way. Draw some pictures of the different types of people you have identified who are most likely to buy jewellery in a wildlife park souvenir shop.

CUSTOMER RESEARCH QUESTIONNAIRE

Please tick appropriate boxes.

Customer: Male ☐ Female ☐

Age group 0-15 yrs ☐ 16-25 yrs ☐ 26-40 yrs ☐ 41 plus ☐

1. Have you ever purchased a piece of jewellery?
   Yes ☐ No ☐
2. For whom did you buy it?
   Yourself ☐ Someone else ☐
3. Are you influenced by current fashion?
   Yes ☐ No ☐
4. What type of jewellery do you prefer?
   (Please circle as many as applicable).

   Gold / Silver / Semi-precious / Fashion jewellery
   Large pieces / Small pieces
   Necklaces / Earrings / Bracelets / Hairslides

'All that glitters is not gold!'. These are made from brass or aluminium with glass (paste) fake stones.

Materials

Consider the properties and characteristics of available materials.

- ▷ Which will have the most interesting appearance?
- ▷ Which will be the strongest?
- ▷ Which will be easy to work with?
- ▷ Which might be combined together in interesting ways?

Look at how craftspeople from other cultures combine materials in unexpected ways.

ICT

Use the Internet and CD-ROMs to help obtain images.
Use a word processor to prepare your questionnaire, and a spreadsheet to show your findings graphically.

Clip brooch in sapphire, diamond and enamel

IN YOUR PROJECT

- ► Use the information and ideas you have discovered to help design your jewellery for the market you have identified.
- ► Always keep in mind the type of customer you wish to attract to buy your product.

Perfect Proportions

The patterns and imagery applied will affect the balance of the design. Will they be two dimensional shapes or three dimensional forms? Will the images be large or small, realistic or abstract?

In Proportion

Natural objects appear to be in perfect **proportion**. We can tell if a made object looks top heavy, unbalanced or 'out of proportion'. Achieving a good visual balance is essential when designing products.

■ ACTIVITY

Sketch free-hand several natural objects. Analyse their design, natural balance and proportion.

Can you find examples which demonstrate the Golden Section or Fibonacci series?

The Golden Section
Artists and designers have been searching for the answer to perfect proportion for centuries.

The Ancient Greeks developed a mathematical solution called the '**Golden Section**', based upon a rectangle with a proportion of 1 to 1.618. This is often simplified to 1:1.6. The Greeks thought that this was a perfect proportion and used it to work out the height and width of their buildings.

This proportion of height to width was also used by Georgian architects when they designed the size of windows. It is also used for working out the size of canvasses for paintings.

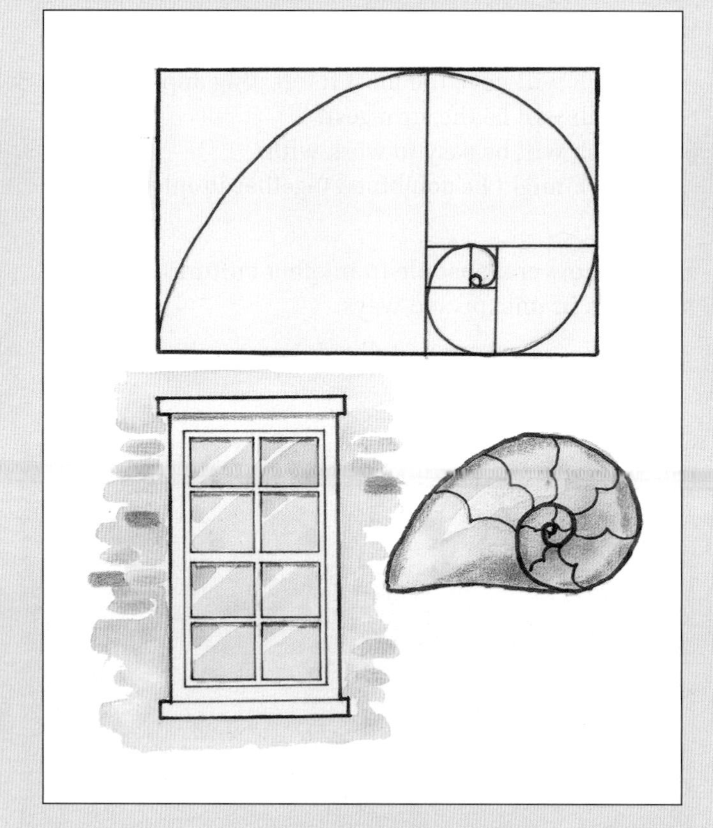

The Fibonacci Series
Leonard Fibonacci (1170 – 1250), the Italian mathematician, devised a formula known as the **Fibonacci series**: 0, 1, 1, 2, 3, 5, 8, 13, 21, 34, 55, 89, 144, etc.

Can you recognise in this sequence of numbers how he developed this formula?

Each number is the sum of the two previous numbers. Choose any number in the series, divide it by the previous smaller number and the answer will be approximately 1.618.

This sequence is often found in nature, for example in the stems of flowers and branches of trees.

Shape and Form

Geometric shapes can provide an excellent starting point for designing objects. How many examples can you see in your surroundings? Try to look beyond the square, circle and triangle. Sketch a selection of geometric shapes, experiment with combinations of shapes, tessellation, cut and shift, mirror images, distortion. Consider natural shapes and combine natural and geometric forms.

ICT

Use a graphics package to experiment with repeat patterns and shapes and colour variations.

IN YOUR PROJECT

Use a variety of presentation techniques to develop your ideas:

- Simple line drawings.
- Shaded drawings.
- Shapes cut out of coloured paper or metallic card.
- Place a textured surface under your drawing and produce a 'rubbing' to simulate textured materials.
- Light pencil crayon shading behind your drawings can be used to highlight the ideas you wish to develop.

Start to think about the appearance of your final product. The shape will be influenced by the part of the body you wish to decorate. The materials you choose will also influence the shape and colour.

As your ideas progress and you experiment with different forms you will narrow down the ideas.

Make several card models of your ideas to try out how they look in position.

KEY POINTS

- Visual balance is essential to good design.
- Geometric shapes and natural forms provide a good starting point for developing ideas. They can be effectively combined to give interesting and balanced designs.

Human Factors

As well as being pleasant to look at, jewellery must be comfortable to wear and securely attached. To make sure your design will be the right size and shape you will need to find out the average proportions of your potential customers.

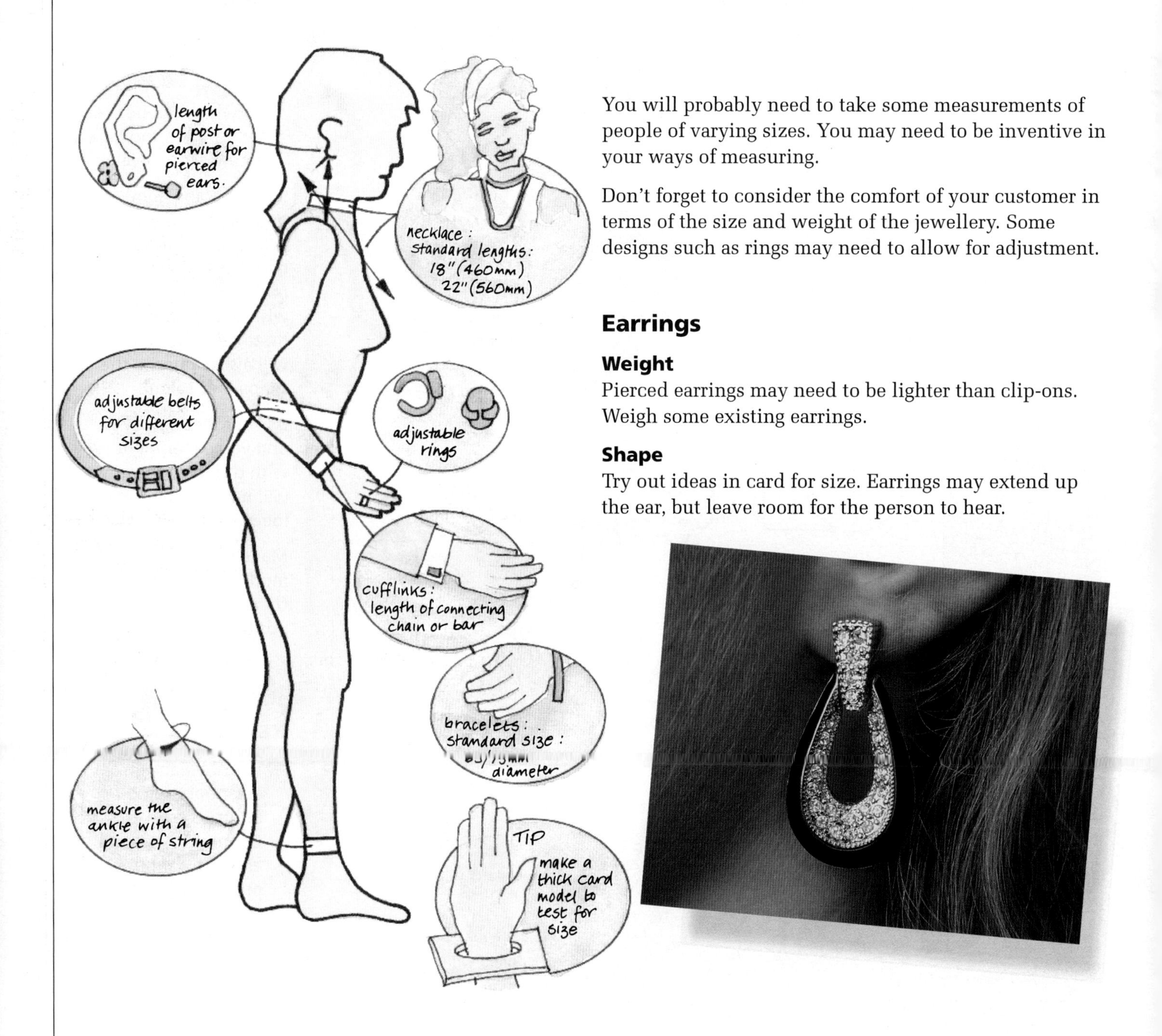

You will probably need to take some measurements of people of varying sizes. You may need to be inventive in your ways of measuring.

Don't forget to consider the comfort of your customer in terms of the size and weight of the jewellery. Some designs such as rings may need to allow for adjustment.

Earrings

Weight

Pierced earrings may need to be lighter than clip-ons. Weigh some existing earrings.

Shape

Try out ideas in card for size. Earrings may extend up the ear, but leave room for the person to hear.

Necklaces

Length

Use string to measure your ideal length. There are also standard sizes : 460mm (18") and 560mm (22").

Bracelets

Size

Round bangles need to slip over the hand comfortably but not slip off when being worn. There are standard sizes: 65 to 75mm diameter. Make a card model to test your design.

Brooches

Weight

They must not be too heavy. The position of the clasp is important to allow for balance.

Rings

There is a range of standard sizes for rings. Each size has a letter code. You will not be able to produce a full range of sizes so, if possible, make an adjustable ring with an open back.

To find out the size of your finger, slip an existing ring that fits you over a ring triblet or a tapered metal bar. Twist wire around the bar in the same position. Cut the wire. Open it out to give the metal length.

IN YOUR PROJECT

- Which measurements will you need to investigate to assist you in the design of your product?
- Can you think of any other important considerations to do with size, shape and comfortable use?

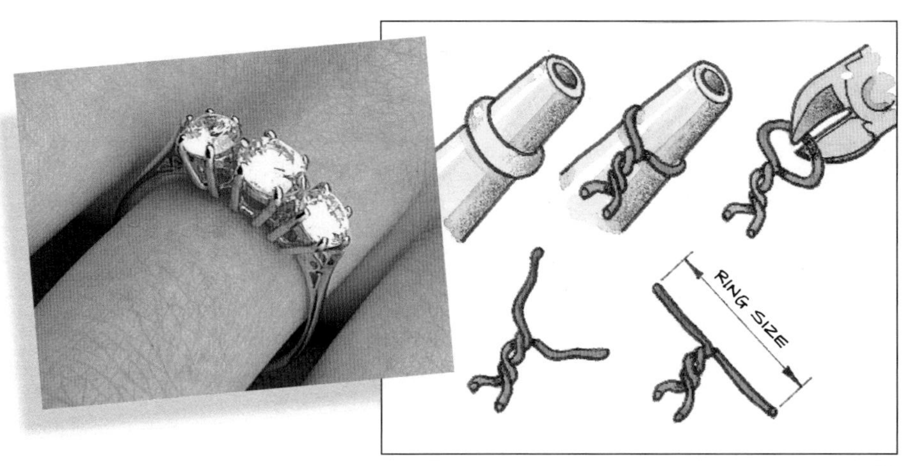

Material Ideas [1]

Traditionally gold and silver have been the favourite materials used for jewellery. Fashion jewellery is designed to reflect current trends, rather than last a life-time. Cheap, light-weight materials such as plastics and aluminium are easy to colour and process in the quantities that the market demands.

www.

To look at examples of contemporary jewellery go to: **www.craftscouncil.org.uk**

Dyed and folded aluminium

Gold and Silver

Gold and **silver** are valuable. They resist corrosion, and temperature changes and are unlikely to irritate the skin. Most importantly they have an attractive feel and colour.

In their natural state gold and silver are too soft to be made into wearable items. It is necessary to produce an alloy with other materials to strengthen them. Gold is alloyed with silver or copper. Silver is alloyed with copper. (See page 130 on alloys.) All gold and silver items are tested at an official Assay Office and a hallmark indicating the quality of the metal is stamped onto each individual item.

Rings, earrings and cufflinks in sterling silver and enamel

Hallmarking

The first **hallmarking** laws were passed in 1238 for products made in London. By 1300 the Act was extended to include the whole of Britain. It remained in force until 1856. The earliest quality marks date from Roman Times!

Current regulations ensure that every article made from gold or silver bears a hallmark which includes:

- a sponsors mark (the manufacturer's mark)
- the standard mark of quality
- the assay office mark
- a date letter of the year of manufacture.

Find an item of precious jewellery. See if you can find out where and when it was manufactured.

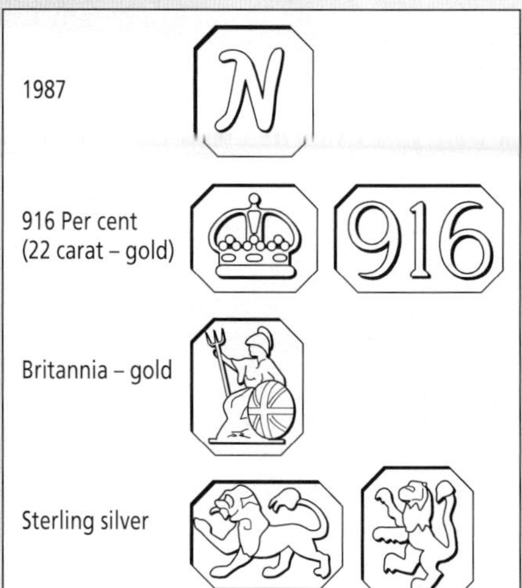

Copper, Brass and Aluminium

Copper, **brass** and **aluminium** are ideal materials for the manufacture of fashion jewellery.

Material	Form available	Characteristics	Uses
COPPER		Pinkish brown in colour. Easily shaped and soldered, conducts heat and electricity well	Pipes, electrical wires, roof coverings
BRASS		Gold in colour. Alloy of copper and zinc. Hard, attractive when polished	Ornaments, pipes, castings
ALUMINIUM		Silver-grey in colour. Light, soft, conducts heat and electricity well. Easy to work	Saucepans, drink cans, cooking foil, aircraft

IN YOUR PROJECT

- Your choice of materials will be influenced by what is readily available in your department.
- Ask your teacher which process you might use to manufacture your jewellery on a small scale.
- How will you combine or join different materials?
- Where could you acquire other unusual materials?
- Consider the weight and working properties, does it need to be flexible, rigid or malleable?
- The cost will be an important factor – gold might not be the best choice!

KEY POINTS

- Gold and silver are traditional materials for jewellers to use
- Fashion jewellery can be made from a huge range of different materials.

ACTIVITY

Draw two items each made from copper, brass and aluminium found in your home or school. Suggest how they might have been manufactured.

Find out more about the process of enamelling. What facilities does your school have for this?

Woods and Plastics

You may wish to exploit the natural qualities of found **wood** such as bark or driftwood.

Commercially available woods offer other possibilitiess. See pages 134 to 135.

The most suitable plastics available to you are acrylic and vacuum forming thermoplastics. See pages 132 to 133

Dyed, laminated and veneered bracelet

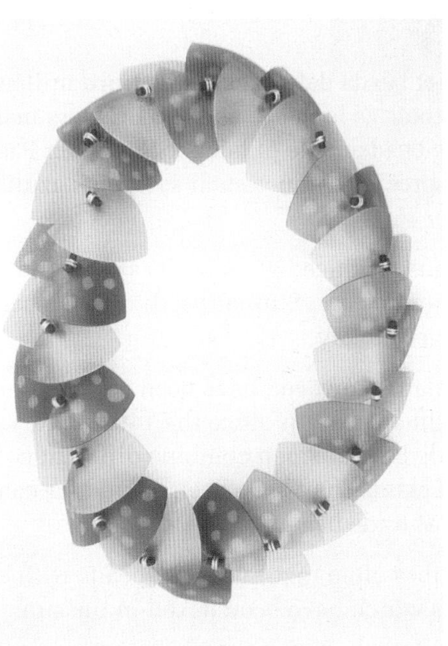

Nylon and brass wire necklace

Material Ideas [2]

For centuries it has been traditional for people to express their status, wealth and life-style by wearing valuable jewellery. Nowadays people wear modern jewellery to make a statement about their individuality. They may want to show they have a sense of humour, or like to be unconventional. These items do not need to be expensive.

What unconventional materials can you think of that might be appropriate for your jewellery?

Make several lists of materials under headings.

Natural and found
Shells
Driftwood
Feathers
Bone
Semi-precious stones
Pebbles
Amber

Natural (manufactured)
Beads
Cane
Papier mâché
Raffia

Man-made
Self hardening modelling clay
Pewter
Fabric
Sequins

Unusual
Drawing pins
Map pins
Rubber tubing
Electrical wires
Computer components

Beads

The earliest beads date back to the third millennium BC. The Egyptians as long ago as 1400 BC were mass-producing beads made of faience (Egyptian Paste) to represent precious stones such as Lapis Lazuli and Turquoise.

The Romans produced glass beads as substitutes for emeralds and pearls. Similar beads are still being manufactured today.

The Venetians, however, have been making the world's most sought after beads since the 14th century. Millefiori beads made by craftsmen combine different coloured threads of glass. They have been valued for centuries and used as trade beads.

African tribes often mix European beads with decorated clay beads which have been baked in the sun.

Nowadays a vast selection of beads from all over the world is readily available from specialist bead shops.

Found Materials

Innovative contemporary jewellery designers have found limitless sources of inspiration from the materials found in their environment, from shells, feathers, pine-cones, gnarled wood to aluminium can ring-pulls, buttons, plastic toys, watch parts and rubber gaskets!

In the 1970s it was fashionable to wear safety-pins in your nose and ears!

Tribal jewellers in Africa produce beautiful jewellery from ostrich shell, horn, bone and even polished seeds.

'Working with recycled materials is an inspiring adventure into the unknown. Because one is working with an unconventional material, in an unconventional way, with unconventional tools, the end result can be very exciting, rewarding and unexpected.'

Val Hunt, Jewellery Designer

Flower Power

Although there are no surviving examples, it is quite possible that the earliest form of body adornment was with flowers and fruits – themes that are still popular today.

Ancient Egyptians wore collars of flowers, olive leaves and berries as an alternative to the heavier more ornate pieces often displayed for ceremonial rituals.

Modern day brides continue the fashion for wearing elaborate headdresses made from beautiful arrangements of fresh and silk flowers.

KEY POINTS

- In the past, a huge range of materials have been used to make jewellery.
- Designers have always used the natural world as a source of inspiration.

IN YOUR PROJECT

What unusual materials could you incorporate into your design?

Shaping and Forming Metals

Metal working techniques have changed little over the centuries. Simple hammered surfaces, stamped decoration, coiled wire, castings and chemical processes to colour metal are all effective alternatives to a highly polished surface.

Tools and Materials

You will need a few basic tools to get you started:

- tin snips or shears
- round nose pliers
- flat nose pliers
- planishing hammer
- old hammer (to add textures to the surface)
- twist drill.

You will also need a selection of material samples such as copper, brass and aluminium sheet and wire. Experiment with some of the ideas shown on this page. Don't forget to keep notes and sketches of how you developed your ideas.

Annealing

Ancient goldsmiths discovered that flexibility could be restored to metallic materials through **annealing**. This involves heating the metal until it is a dull cherry red (not bright orange!). This process transforms the metal's grain structure, making it soft and workable. It will gradually harden again as you work it.

A fine layer of oxide will form on the surface. You may wish to exploit this feature in your jewellery or it can be removed by soaking it in an alum solution or 'pickle'. Wash under water and dry before you start work.

Methods of Forming Wire

Joining rings

Wind the wire around a rod, slide off the coil and use tin snips or a piercing saw to cut off individual rings.

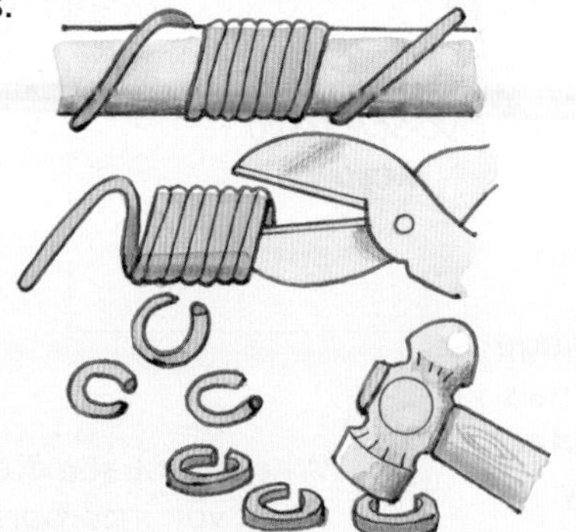

To produce flat rings hammer on an old surface plate or anvil.

Coiling wire

1. Bend up a small ring using round nose pliers.
2. Hold the ring flat in a pair of flat nosed pliers. Rotate the wire tightly to produce a coil.

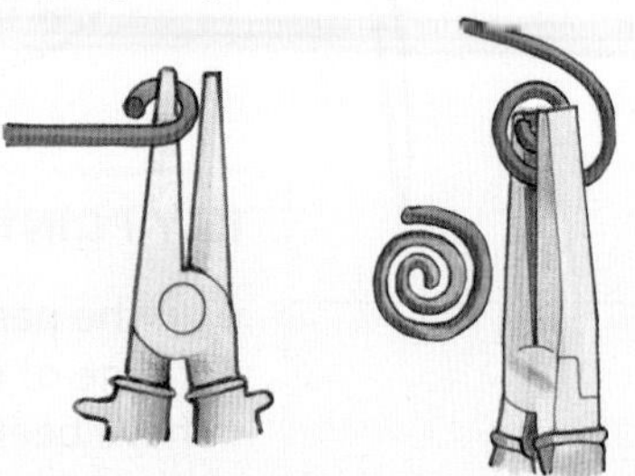

Use flat nosed pliers to produce zig zags and other shapes.

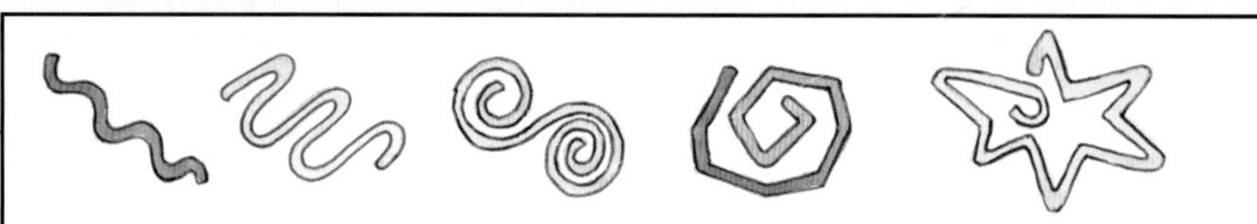

Twisting wire

Two or more types of an annealed wire may be twisted together.

1. Clamp one end of the wires in a vice.
2. Put the other ends securely in a hand drill.
3. Hold the wires taught and twist to the required tightness.

Sheet Materials

Hammered finishes

1. Make sure your metal has no sharp edges.
2. Anneal to soften (not necessary for aluminium).
3. Place the metal on an anvil.
4. Small round hammer marks can be obtained by using a planishing hammer. Take care to place each mark and work from the centre outwards.

Textured effects can be obtained by filing or grinding or drilling small holes in the face of an old hammer.

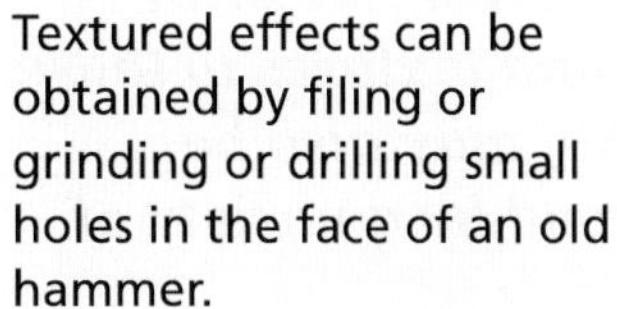

Melting the surface

Heat the metal until the surface just begins to melt – remove the flame immediately!

This process is usually carried out before the metal is cut to shape

Roll texturing

If your school has a set of rolling mills there are a variety of different effects that you can achieve.

Place metal mesh or wire between the two sheets of annealed metal, to form a sandwich.

Pass through the rolling mill. An impression will be formed on both surfaces, facing the centre.

SAFETY FIRST!

- Hold small items with pliers when hammering
- Wear safety goggles and protective leather apron when using the forge.

Impressed Design

To produce simple **impressed designs** in metal try the following method.

1. Cut out a simple shape in thin copper or brass sheet.

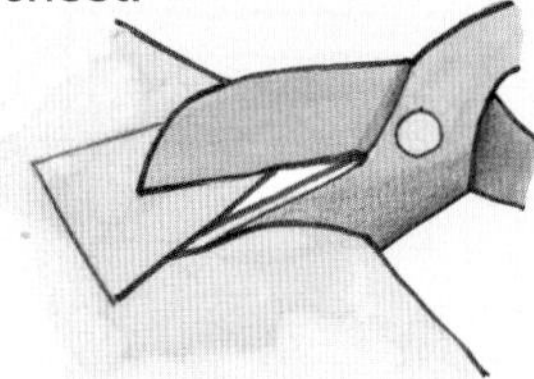

2. Anneal it and leave to cool. You may wish to leave the oxide layer to produce an 'antique' finish or you can soak it in alum or pickle to remove the oxide layer.

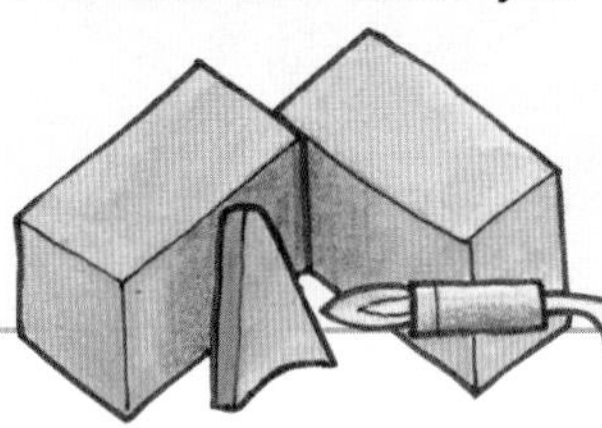

3. Bend unannealed brass wire into simple shapes (see bending wire).

4. Tape your wire into position using two layers of masking tape. Tape both sides to protect the surface.

5. Place wire side down onto an old anvil. Carefully hammer until an impression of your pattern appears in the masking tape.

6. Remove tape and wire.
7. Polish or leave oxide finish.

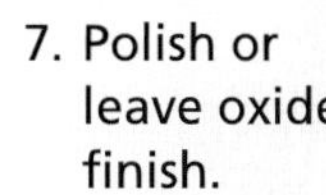

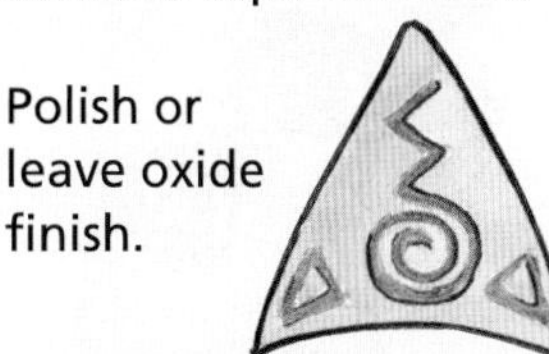

This method may be used for small batch production. Replace the brass wire with steel as it is harder wearing.

IN YOUR PROJECT

- There are endless ways in which natural and man-made materials can be combined, manipulated and processed. Many successful products are the result of inspiration drawn from experimentation with the properties of materials.
- The most attractive designs are often produced using very simple techniques and some imagination.

Texturing Acrylic and Wood

Acrylic sheet usually has a hard flat shiny surface. Similar patterns and textures to those used for metal (see page 59) may also be achieved in plastic and wood by a variety of methods.

Press-forming Acrylic

Stamping blocks

Patterns and textures may be indented into acrylic by using **press-forming** stamping blocks. You will need:

- 2 blocks of MDF approximately 70 × 70 mm, minimum thickness 10 mm
- several assorted squares of acrylic sheet 50 × 50 mm, opaque or transparent, including some with light reflecting properties
- copper wire
- 2 × 12 mm panel pins
- Twist drill, pliers, tinsnips.

To make your press-forming stamping block follow the instructions below:

1. Draw around one of the acrylic squares onto one of the MDF blocks.

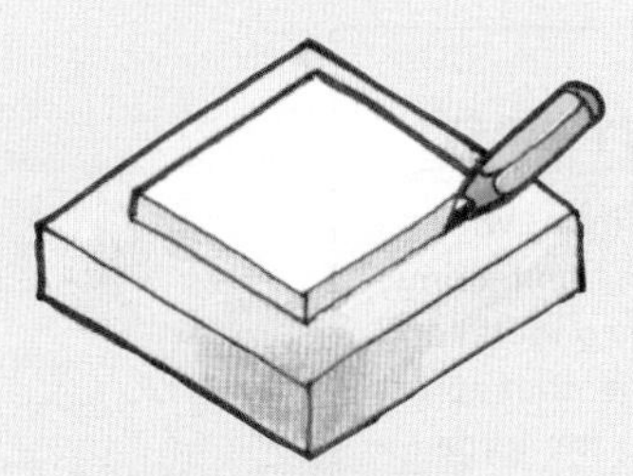

2. Bend up copper wire flat shapes. Super-glue into position, within the pencil outline. Drill small holes to add variety in the pattern.

3. Hammer the panel pins into the decorated block, leaving them protruding approximately 10 mm. These will provide location points for the acrylic sheet.

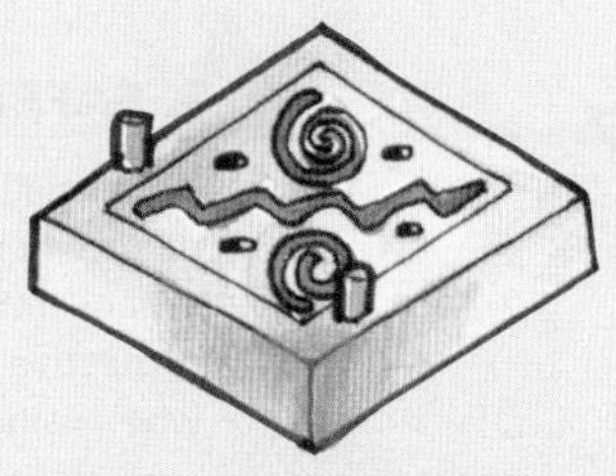

4. Using a drill bit slightly larger than the diameter of the panel pins, drill through one of the other blocks of MDF to position holes for the location points.

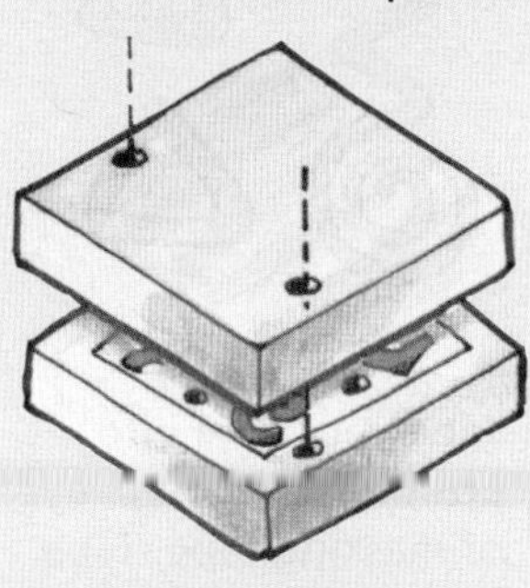

5. Place the acrylic sheet into an oven set to 160 °C. Heat the sheet until it goes floppy.

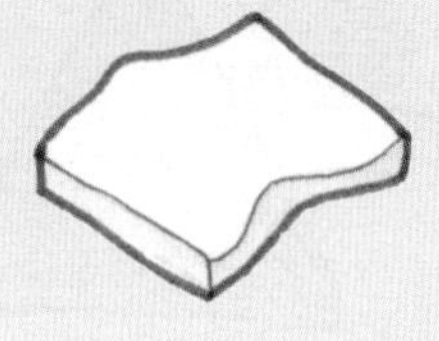

6. Wearing protective gloves, carefully remove the hot sheet and place in position against the protruding panel pins; quickly slot the second block of MDF over the panel pins.

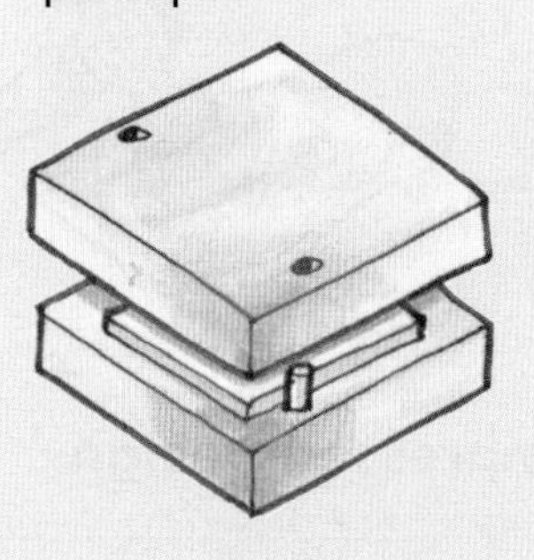

7. Quickly press the two parts in a vice as fast as possible because the sheet will soon harden.

8. Leave to cool for a few minutes, remove from the vice and lever off the impressed acrylic sheet.

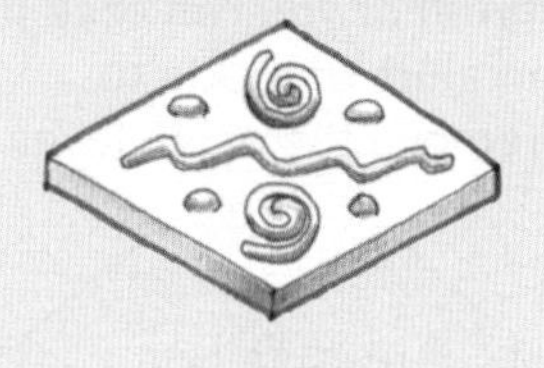

Inlaying Copper Wire and Acrylic Rod

It is possible to inlay copper wire into the surface of acrylic using the method shown. You will need to use only enough super-glue to 'tack' the wire to the forming block, to enable it to be released in place.

Interesting effects can be achieved by **inlaying** clear acrylic rod into clear acrylic sheet and press-forming between two smooth MDF blocks. Cut the rod slightly longer than the thickness of the sheet to achieve a transparent 'ring' around the rod.

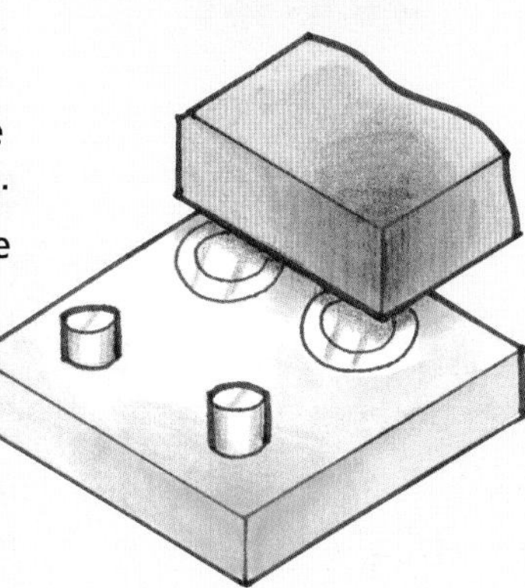

Press-forming Textures

Experiment with pressing different textures into the surface of heated acrylic sheet e.g. coarse sand paper, strips of card, mesh, etc. Try this for yourself and see how many different textures you can achieve.

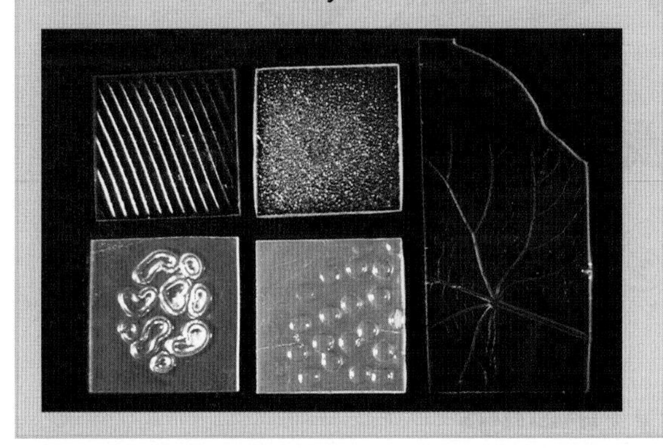

IN YOUR PROJECT

- Remember that there are many ways of **texturing** the material which you have chosen to use. All you need is imagination and the patience to experiment until you get the texture that you want.

Wood and Metal

Interesting effects can be burnt into the surface of wood by a technique called pyrography. It is possible to use a soldering iron to singe the wood or you can bend or file a shape into a long length of mild steel. Heat this in a flame and press into the wood.

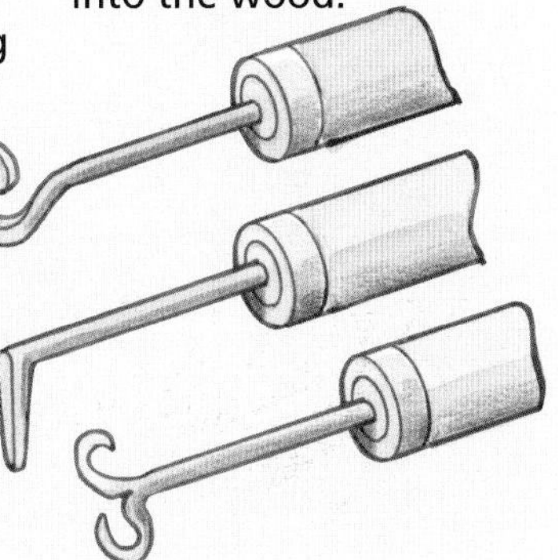

Metal punches can be made by filing patterns into mild steel rod. Hammer the punch onto sheet metal to produce a pattern.

KEY POINTS

- Acrylic must be heated first before textures can be applied.
- Care must be taken when handling hot acrylic .
- Interesting surface patterns can be created by inlaying and punching.

Joining and Fixing Materials

There are many ways to join materials. Consider some of the methods such as* soldering, rivets and pegs *and* jewellery *'findings'.

Soldering

Once the different parts of commercially produced metal jewellery have been produced, they are usually soldered together. This is a highly skilled operation. There are also many occasions when, due to the combination of materials, soldering is not an appropriate joining method.

There are however a number of alternatives to soldering.

Rivets and Pegs

Commercially available pop-rivets could be used to join acrylic and metal, adding an extra dimension to your design.

However you may wish to use a more discrete method, by making your own tiny rivets to join the components together.

1. Drill a hole through all of the sections to be joined.
2. Select a soft metal such as copper wire which fits tightly in the hole.
3. Cut the wire leaving ½mm protruding from each side.
4. Place the parts onto a steel block, carefully tap the copper wire on each side of the hole.

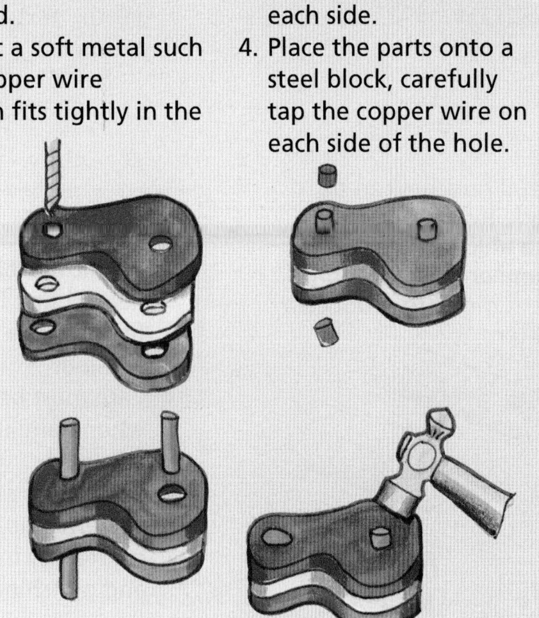

Jewellery Findings

A vast selection of jewellery findings are available. These are basic metal components such as brooch pins, earwires, clips, tie pins and necklace clasps. They will transform your small decorative objects or pile of beads into wearable jewellery.

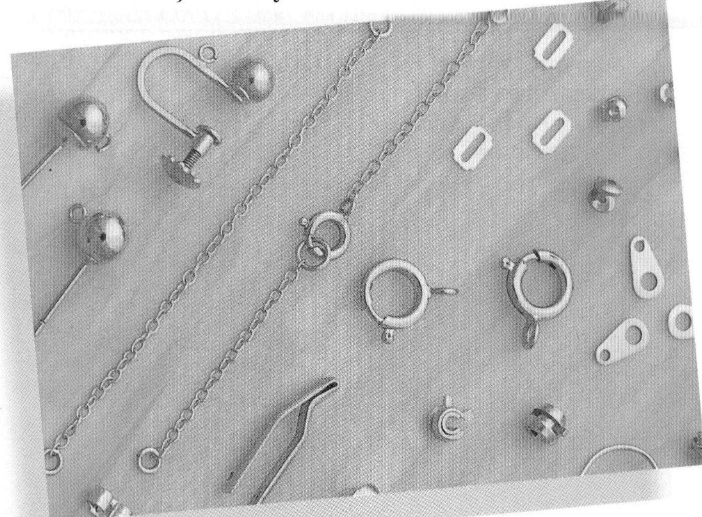

Links and Chains

Simple chains may be produced using the method for making round rings then joining several together. They can look very effective when the links are flattened.

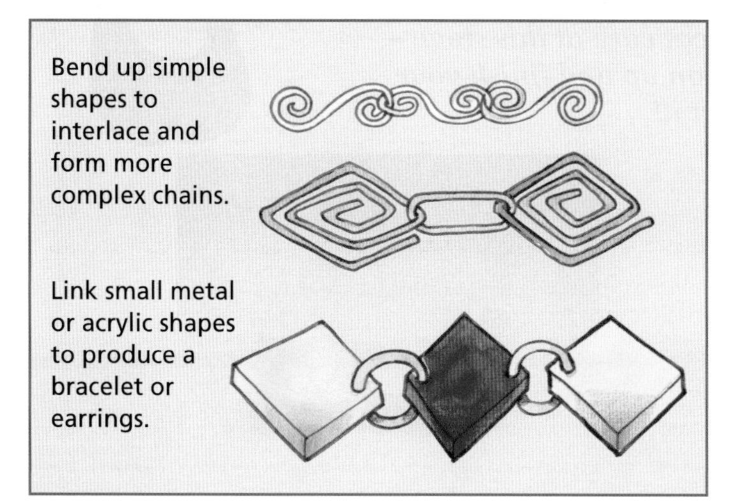

ACTIVITY

Experiment with making your own simple jewellery fittings from wire, e.g.:

- findings
- jump rings
- ear wires – bend your own designs from silver-plated or silver wire. Copper wire is not suitable as it may cause an allergic reaction to the skin.
- necklace clasps, for example hooks and rings.

KEY POINTS

- There are a wide variety of ways of joining materials. These include ready-made components and those you can make yourself.
- Adhesives must be carefully selected according to the materials that need to be joined. It is important to follow the manufacturer's instructions carefully.

IN YOUR PROJECT

- Find out which commercially produced findings are available to you.
- Will you need to produce your own jewellery findings?

Adhesives

There are many varieties of glues or **adhesives** available. Select the correct adhesive for your job.

- Epoxy resin (e.g. Araldite Rapid): an excellent adhesive which will join most materials effectively. It consists of two parts – an adhesive resin and a hardener. They have to be mixed together in equal amounts.
- Tensol cement: suitable for joining acrylics (see page 102).
- Latex adhesive (e.g. Copydex): suitable for joining paper, fabrics and found materials.

Gluing tips

1. Ensure that both surfaces are free from grease, dirt or moisture.
2. Lightly scratch the surfaces to ensure a good contact area, prior to fixing.
3. Be prepared – you may need small clamps or masking tape to secure the joint.

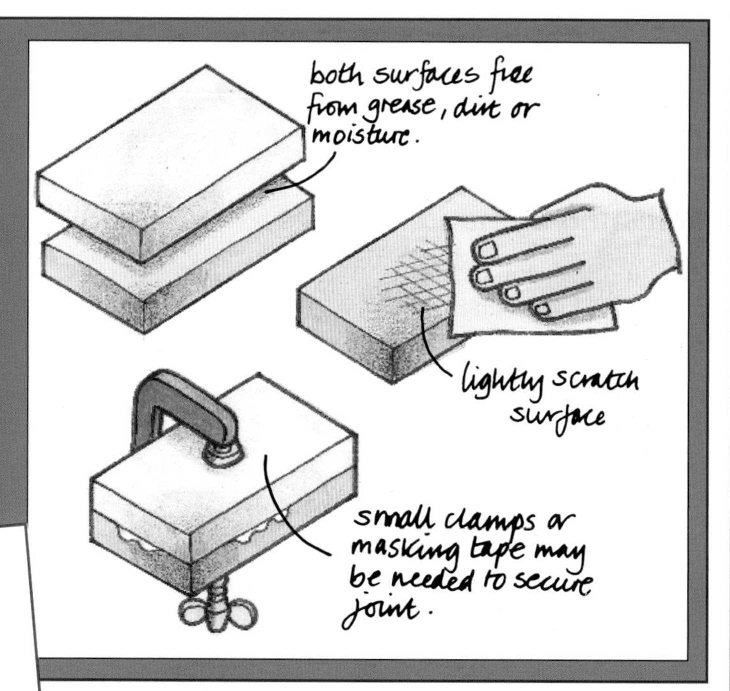

SAFETY FIRST!

- Always read the manufacturers' instructions.
- Many modern adhesives contain solvents which may be addictive or harmful. Use a well-ventilated area.
- Avoid contact with the skin.

Finishing Off

The polish and finish you apply to your jewellery in the final stages of making can transform a rough piece of metal work into a beautiful object. Take great care at this stage – it may take as long to clean up and finish your work as it did to construct it!

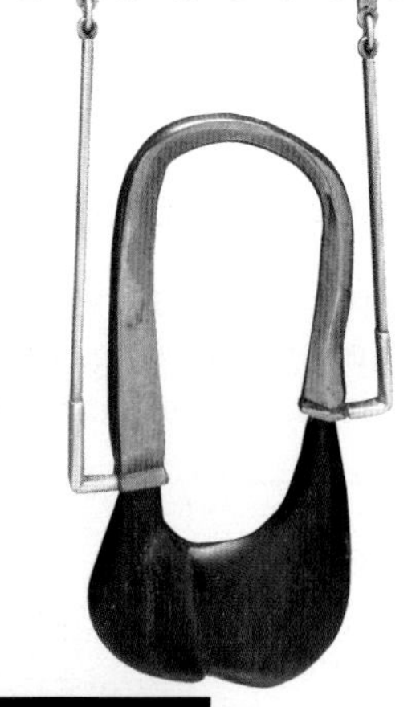

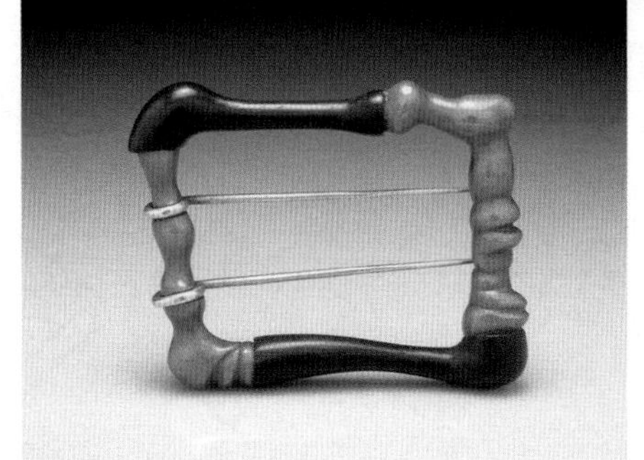

Preparation for Polishing Metal

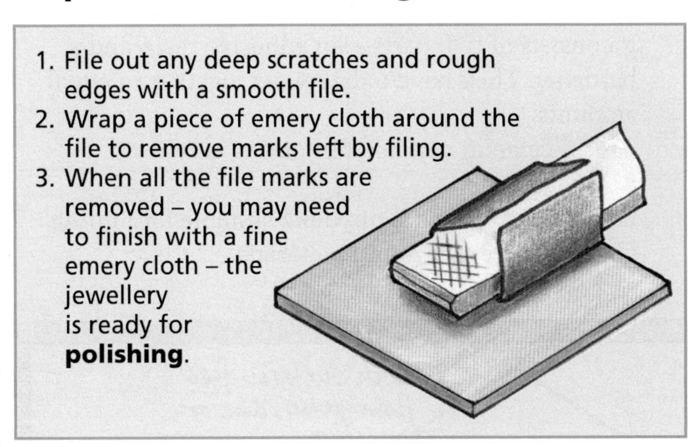

1. File out any deep scratches and rough edges with a smooth file.
2. Wrap a piece of emery cloth around the file to remove marks left by filing.
3. When all the file marks are removed – you may need to finish with a fine emery cloth – the jewellery is ready for **polishing**.

Polishing by Hand

A high polish can be obtained by using a **buff stick** – a simple tool you can make.

Making a buff stick

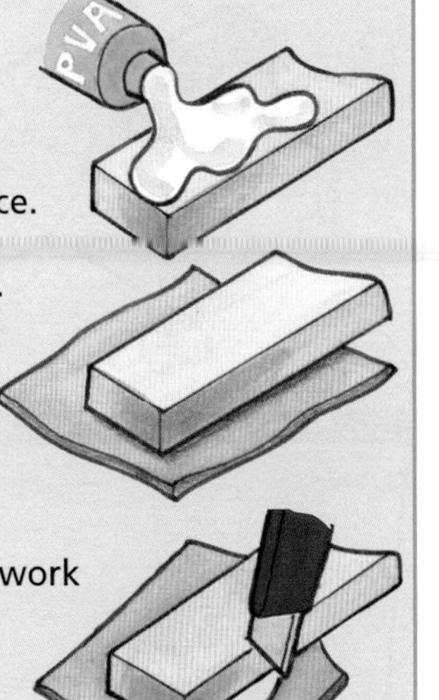

1. Select a suitable piece of wood approximately 200 × 20 × 10 mm.
2. Apply PVA glue to one surface.
3. Place glue side down onto a piece of suede – leave to dry.
4. Trim the edges of the suede using a craft knife – your buff stick is now ready for use.
5. Apply Tripoli polishing compound onto the suede.
6. Rub the buff stick over your work to gradually remove any scratching and produce a shine.This is a slow process and will take quite a bit of effort to produce a bright shine!

Threading

This is used to polish inside small holes

1. Tie several strands of soft string (300 mm long) to a hook on the bench (a shoe lace will do!).
2. Apply Tripoli polishing compound to the string.
3. Thread the string through the hole.
4. Hold the string taught and rub the jewellery up and down the string.

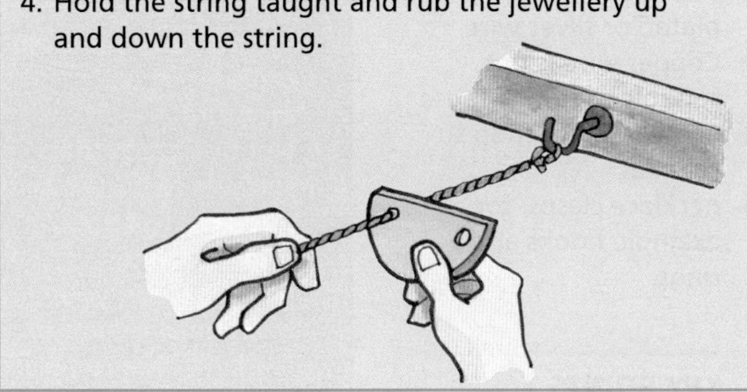

Burnishing

A bright shiny edge can be obtained by rubbing firmly with a burnisher to compress the surface after finishing with emery cloth.

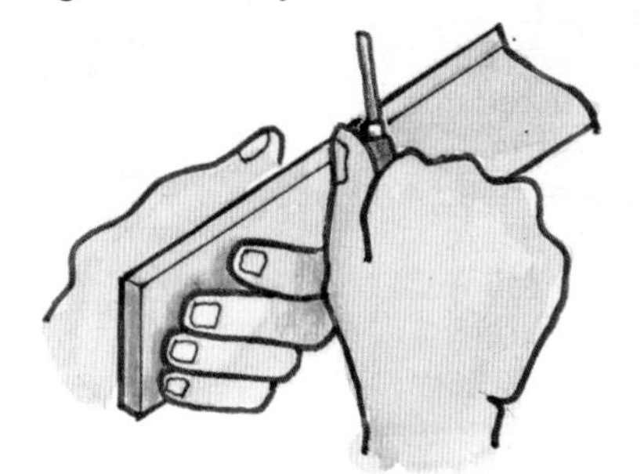

Buffing Machine

If you require a bright, polished, shiny surface you can use a **buffing** machine. Great care needs to be taken. It is not suitable for small items – never polish a ring on your finger. You must ask your teacher for advice on how to use the machine safely.

Tips for safe buffing

1. Tie hair back, secure loose clothing and wear protective glasses.
2. Stand back from the mop, switch on the machine.
3. Carefully press the Tripoli polishing compound against the rotating mop.
4. Firmly hold the jewellery against the mop, just below the centre spindle – if you hold it too tight it may catch against the mop and be thrown from your hands.
5. Keep the jewellery moving against the mop until all surfaces are polished.
6. Don't forget to turn off the machine when you have finished.
7. Wash the jewellery with detergent to remove excess Tripoli polish.

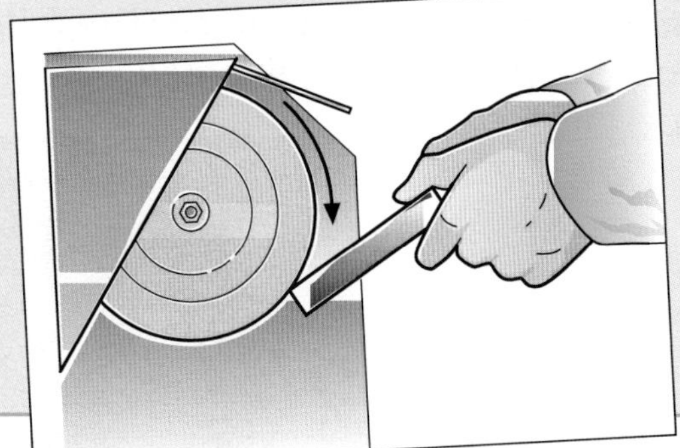

Note: Do not attempt to polish chains on a buffing machine as this is extremely dangerous. The chain may catch on the mop and drag your hand into the machine.

Colouring Metals

Jewellers have traditionally introduced colour into their designs either by including precious stones or combining different metals. More recently electrical processes have been developed to produce a vast spectrum of coloured oxide finishes on aluminium, titanium and niobium. It is possible to produce several attractive colours on copper by simply applying heat.

When heat is applied to a metal, the surface atoms combine with oxygen and cause a layer of oxide to form. As the surface gets hotter the oxide will proceed to change colour. It is possible to preserve these colours through cooling.

Colours may be preserved at any time by quenching in engine oil. The oil will deepen the colour.

1. Place a piece of copper sheet against a fire brick on the forge. Gently apply a flame. Watch the surface colours change.
2. Continue to heat, it will form a red cuprous oxide layer.
3. If you add more heat the oxide layer will thicken and toughen to produce a black cuprous oxide layer.
4. Eventually the oxide may peal to reveal the red oxide layer below the surface.

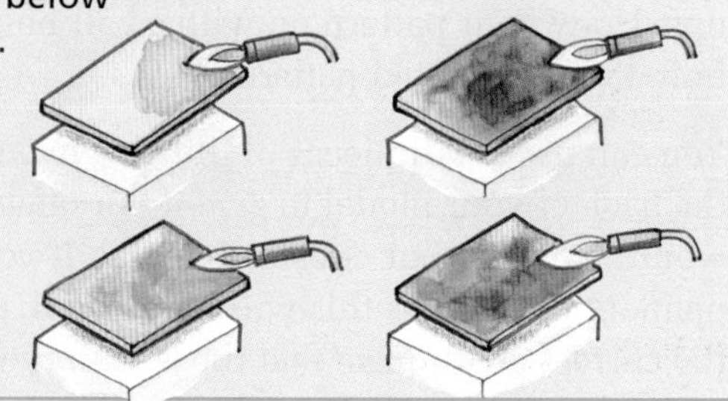

■ ACTIVITY

Try out some experiments with scrap copper sheet. Try heating the copper for different lengths of time and see what effects you can produce. Don't forget to keep notes and sketches explaining how you achieved different finishes.

Mask out areas with masking tape or polish away flat surfaces to reveal the oxide layer in depressions caused by the hammering and wire technique.

How could you combine different finishes using some of the joining techniques mentioned previously?

Blackening Copper or Silver

It is possible to obtain a black surface on copper or silver by applying potassium sulphide. You will need to ask your science department to supply the chemical and allow you to use the fume cupboard to extract the unpleasant smelling fumes.

1. Dip a nylon paintbrush in warm water.
2. Load the brush with potassium sulphide (in the same way as using block paints).
3. Paint the areas you wish to darken e.g. the recesses caused by stamping.
4. Leave to dry. Wash your hands.
5. Polish the jewellery with a buffing machine to reveal the black indentations.

Finalising the Design

Remember to refer back to the customer specification to ensure you have answered the design brief. Select a design that will be both attractive and economical to produce. You will need to devise a method of small batch production.

What Will it Look Like?

When you have selected your best ideas, make several **mock-ups** or **prototypes** to test your designs for size and appearance. You may find that your ideas look completely different when they are transformed into 3D and it is a good idea to check before you make mistakes with expensive materials.

Card with a metallic finish is excellent for making mock-ups. Draw your pattern on with a ball point pen to imitate the indented patterns.

You can use small pieces of Blu-tack evenly spaced over the back of your model to give it the same 'weight' as it would have in your chosen material. If you are making a pendant or earrings this will help you to get the holes in the correct position so that the jewellery will hang properly. Pin, tape, suspend or hold the models in position to gain an idea of how they will look when worn.

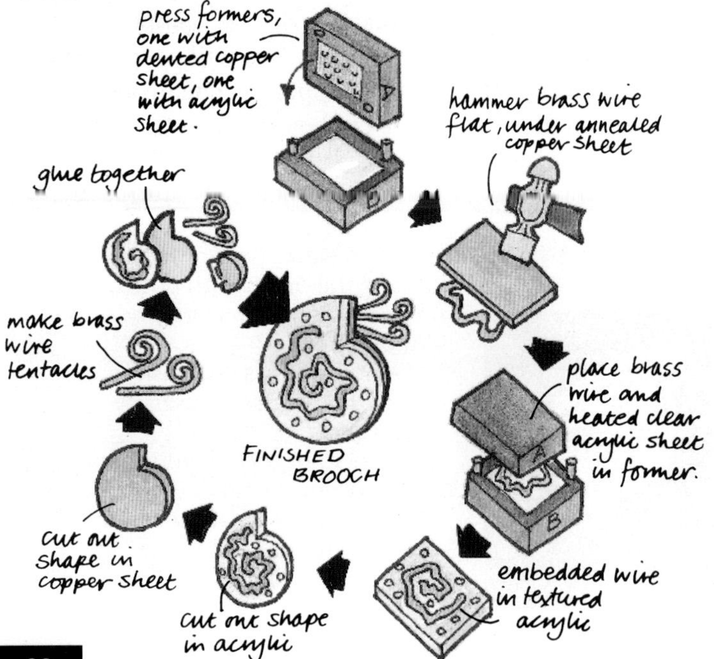

Try out your ideas against photographs from magazines

Planning the Making

You will need to:

- ▷ plan the sequence of tasks and procedures for making, identifying all the main stages in the manufacturing process.
- ▷ produce a 'cutting list' which includes details of materials and components needed – material, thickness, dimensions, length of wire, etc. This should be done in the form of a matrix.

Remember to:

- ▷ include materials for any tools which you may need to make, for example an MDF pressing tool.
- ▷ work out the best arrangement of your parts on acrylic or metal sheet to avoid wastage.
- ▷ list the tools and equipment you will need.

Working Drawings

When you have chosen your final solution you will need to produce a **working drawing**. It may not be necessary to present your idea as a three-view orthographic drawing, although you could produce a plan and front elevation of your jewellery. The important thing is to include all the information which would enable someone else to make your design.

You should include the following details:

- the names of materials to be used
- the tools and equipment which you will use
- a list of parts
- the dimensions
- the method of joining the components
- any special finishes which need to be applied and how to do this
- the fittings or findings required, either commercially available or home-made.

ICT

Word process your plan for making and cutting lists. Use CAD to work out the best use of material when making more than one item.

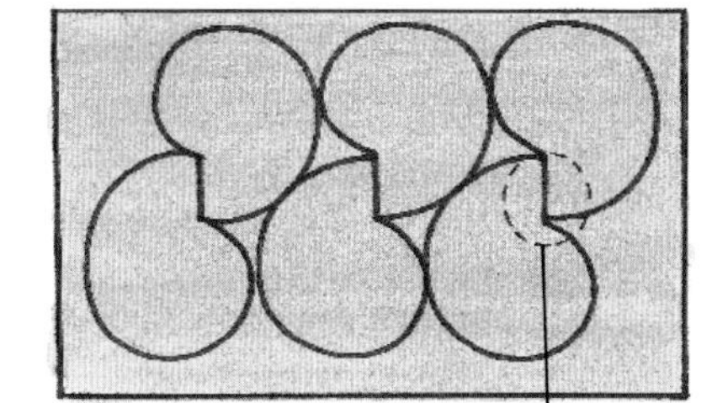

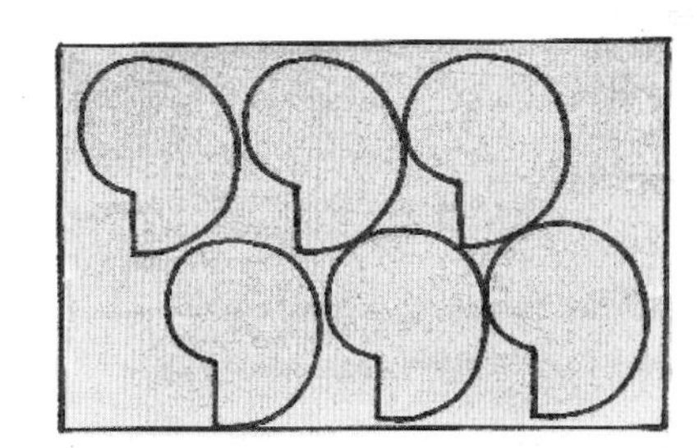

this would be difficult to cut out

inlaid brass wire

large brooch pin

polished finish

copper wire binds brooch together

spots represent animal markings

mass produced clip

brass wire hammered flat

copper sheet shape, oxidised finish

IN YOUR PROJECT

- Gather together all of the information you have collected towards your project. Do not throw anything away – even rough sketches.
- Present your ideas and experiments clearly using sketches, drawings and notes to explain how you gained your knowledge and developed your ideas.
- Remember to include references to books and catalogues which you may have used, either for inspiration or to choose materials and jewellery findings.

Counting the Cost

Manufacturers have to take a wide range of factors into account when working out the production costs of new designs. Keeping the costs down is very important.* Batch production *can help.

Item	Size	Qty	Cost (p)
Pressing tool			
MDF	50 x 50 x 12mm	2	10
Copper wire	15 x 1.5mm		15
One pair of earrings			
Copper sheet	40 x 40 x 1mm	2	40
Copper wire	20 x 1.5mm		20
Acrylic sheet	40 x 40 x 3mm	2	10
Ear fittings			10
Extras, e.g. adhesives			10
TOTAL			1.15

Costing for materials for earrings (all prices are approximate). How much would it cost to make ten pairs?

Money Matters

It is essential to keep **production costs** down to a minimum. Careful planning helps avoid unnecessary wastage of materials and time. Excess materials may be recycled or used in another product.

Materials

The most suitable material may not be the cheapest. Often more expensive materials are more economical to process.

Components

Compare the prices and suitability of components and fittings from several catalogues or suppliers.

A manufacturer would also need to take a further range of costs into account.

Energy

The production costs involved in running machines, such as for electricity, can be expensive, especially if heat-processes are involved.

Labour

This is often the most expensive factor to be considered. In business, the time for designing and making a product is usually costed by the hour.

Premises / capital expenditure

The workspace has to be paid for, and new items of manufacturing equipment may need to be purchased.

Packaging / transport / retail commission / tax

Each item has to be packaged and transported to where it will be sold. A retail outlet will also need to add on their profit (this can range from 15 to 50%) and VAT at 17.5%.

ICT

Use a spreadsheet to help calculate different costings.

IN YOUR PROJECT

Your product will be developed and manufactured in school time, so you may choose not to include your labour cost.

However, you will need to take into account the cost of making any specialist tools that you may need – for example the MDF pressing tool, materials, jewellery findings, colouring chemicals, adhesives and the card on which to present the final product, and any other materials you use.

Batch Production

The more you decide to make, the cheaper it will become. Savings can be made by buying materials and components in larger quantities. Tools and jigs only need to be made once, and can speed the production process up, reducing labour costs.

Work out how much it would cost to obtain the materials and components to produce 1,000 copies of your design. You will need to look through suppliers' catalogues to discover the costs of buying in larger quantities.

Try to avoid specifying more material than you need. It might be more economical to make a slightly smaller number to avoid waste.

Case Study: Featherstone and Durber's Pocket Knives

To find out more about Featherstone and Durber's knives go to: **www.made-in-sheffield.com/F+D/**

Some jewellery designers also design things which are not meant to be worn. Below, Featherstone and Durber describe the distinctive pocket-knives they make.

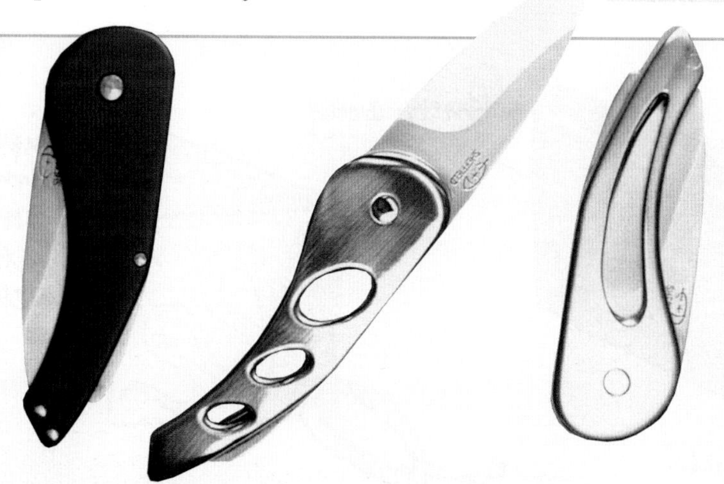

'For many generations Sheffield was at the heart of the steel industry and known around the world for quality tools. By making pocket-knives in Sheffield we are reviving a centuries old craft which was in danger of being permanently lost.

Our knives are manufactured from the finest materials, carefully selected for form and function. We create a unique combination by carefully blending traditional steel with modern elements.

The knives are made to an original design. The ingenious safety locking mechanism ensures that the knife remains securely closed and cannot accidently open in a bag or pocket. In use the blade engages a catch which keeps it rigid and effectively avoids the risk of closure on the fingers.

Every knife is the product of numerous operations and many hours work. The blade is precision stamped from high grade stainless steel using a specially designed tool. It is then drilled for the rivet and hand ground before being hardened and receiving its first finishing. The spine is pressed out, shaped and drilled to take the wire spring. Meanwhile the material for the handle is selected and shaped, before the final assembly and finishing.

Although machinery is used to simplify repetitive tasks it is the skill of the craftsman which makes the knives special. Each knife goes through many processes and at every one it is exhaustively checked for quality. The hand finishing and the use of different materials for the handles mean that, while the quality always remains consistent, each knife has its own individual personality and no two knifes are ever quite the same.

The result is a product to treasure – an artefact that will retain its blend of elegance and functionality over a lifetime of use.

These knives are becoming increasingly popular as corporate gifts. They can be engraved on blade or handle with a company logo. In addition we produce one-off limited edition and custom-made knives. Each one of these is totally hand made throughout and often incorporates decoration in silver or gold.'

Featherstone and Durber

Product Presentation

The* presentation *of a product at the point-of-sale needs to attract a potential customer to your design. Good* display *and* packaging *are important. What kind of image do you wish to promote?

Think carefully about the type of customer you have designed the jewellery for. Design a multi-purpose card which may hold earrings or brooches, tie pins, etc. Keep the design simple – do not let it overpower the actual product. Look at how jewellery is presented for sale in the High Street.

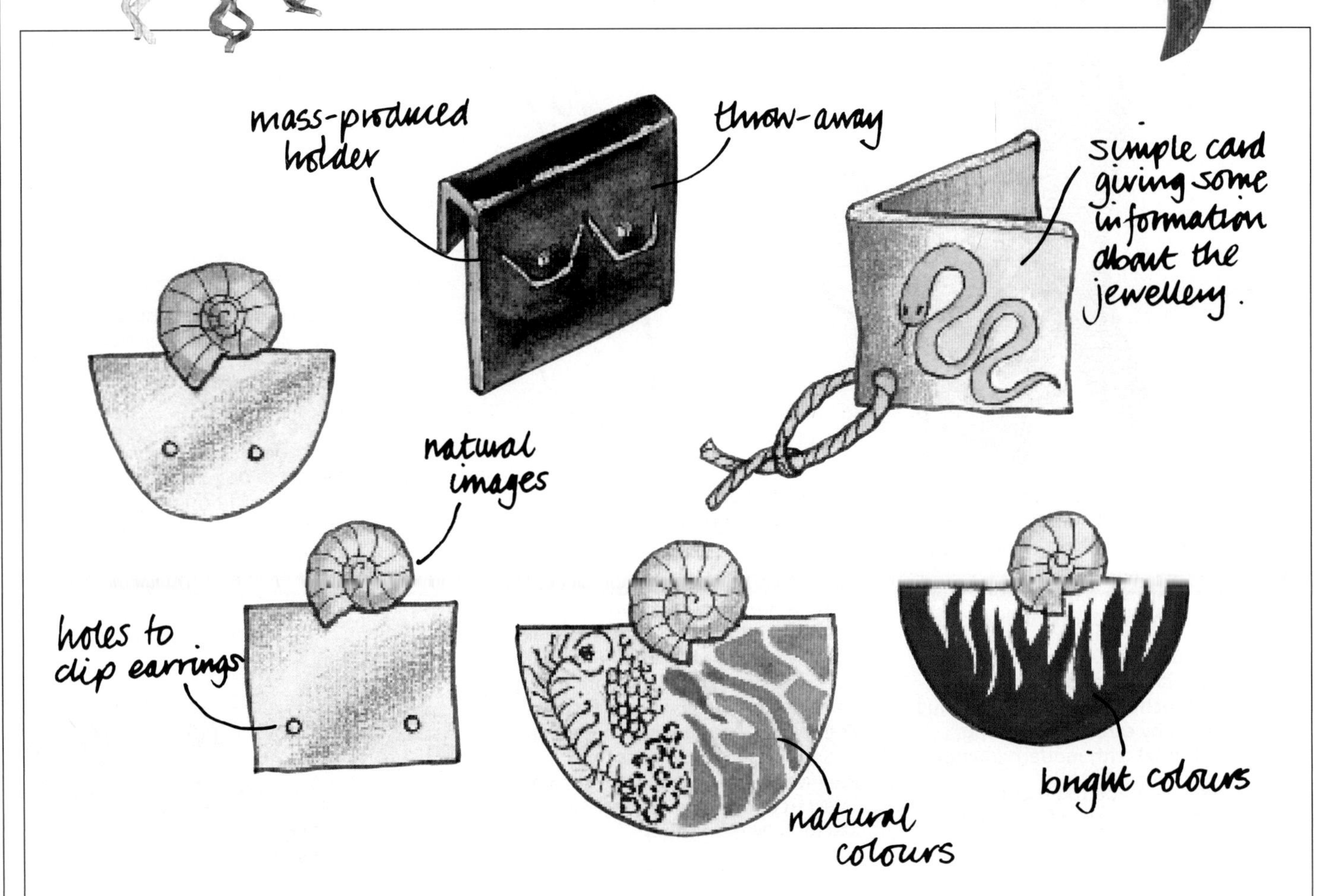

Testing and Final Evaluation

Testing Your Final Product

Designers and manufacturers are constantly trying to improve their products, **testing** ideas, applying quality control, surveying customers to give value for money. You may wish to conduct a survey or ask yourself a few questions about how you might improve your manufacturing processes and final product.

- Is it comfortable to wear?
- Is it too heavy?
- Is it too big or too small?
- Are there any sharp edges?
- Will the applied finishes rub off? (How can you test this without spoiling the product?)
- Were the processes which you used appropriate?
- Were the materials appropriate?
- Do the materials cause a skin reaction when worn next to the skin?
- What do other people think about your product?

WILDLIFE WORLD

PRESS RELEASE • PRESS RELEASE

Wildlife World are pleased to announce the launch of a new range of jewellery for sale in its shops. These attractive hand-made pieces will make an ideal souvenir for the visitor, or presents for family and friends.

Evaluating Your Design Folder and Final Product

Here are some of the key things you need to present in your **design folder**. Check that you have:

- communicated your ideas effectively using sketches and notes
- reviewed your work critically, rejecting unsuitable ideas and explaining why
- used specialist terms and vocabulary where appropriate
- checked your spelling and punctuation
- considered Health and Safety issues.

Here are some key elements of your final product. Check that you have:

- compared your final design with what the client asked for
- compared the quality of your product with the quantity and price required
- tested your final product.

Are you pleased with your final product?

How long did it take to make the jewellery?

How could you improve the quality and timing of your work?

Examination Questions 1

You should spend about one and a half hours answering the following questions. To complete the paper you will need some A4 and plain A3 paper, basic drawing equipment, and colouring materials. You are reminded of the need for good English and clear presentation in your answers.

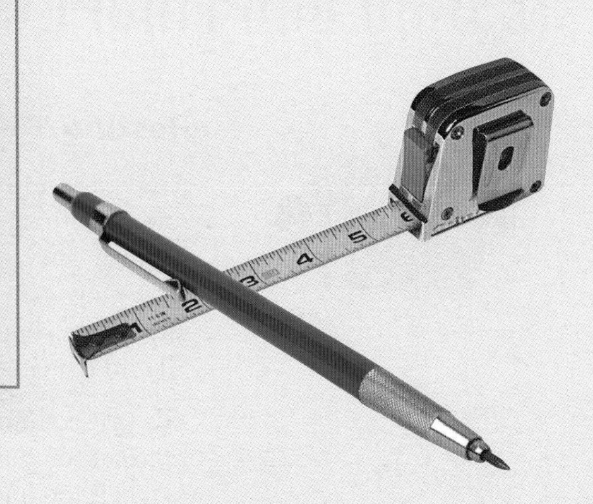

1. This question is about design specification. See pages 16-17. *(Total 5 marks)*

You have been asked to design a clock for a child's bedroom. Write down five things that you will need to think about. *(5 marks)*

2. This question is about developing a design. See pages 36-37. *(Total 6 marks)*

The illustration below shows a very simple idea for a clock.

a) Make your own drawing showing three ways in which the design for the clock can be developed. *(3 marks)*

b) Make notes on your own drawing to show what these changes are and why you have made them. *(3 marks)*

3. This question is about manufacturing. See pages 34-35. *(Total 2 marks)*

The clock will be vacuum formed. Give one reason why this manufacturing method is a good choice for the batch production of small items. *(2 marks)*

4. This question is about characteristics of materials. See pages 40-41. *(Total 3 marks)*

a) Name one material that is suitable for making a former. *(1 mark)*

b) Give two reasons for choosing this material. *(2 marks)*

5. This question is about manufacturing. See pages 34-35. *(Total 14 marks)*

a) Name two materials that are suitable for vacuum forming the clock face. *(2 marks)*

b) Draw the six steps for using a vacuum former. *(12 marks)*

6. This question is about quality assurance. See pages 40-41. *(Total 4 marks)*

After you have finished vacuum forming a prototype, you find that there is 'webbing' on some of the edges.

a) Give two reasons for why this might have happened. *(2 marks)*

b) Write down two things that you do to make sure that it does not happen next time. *(2 marks)*

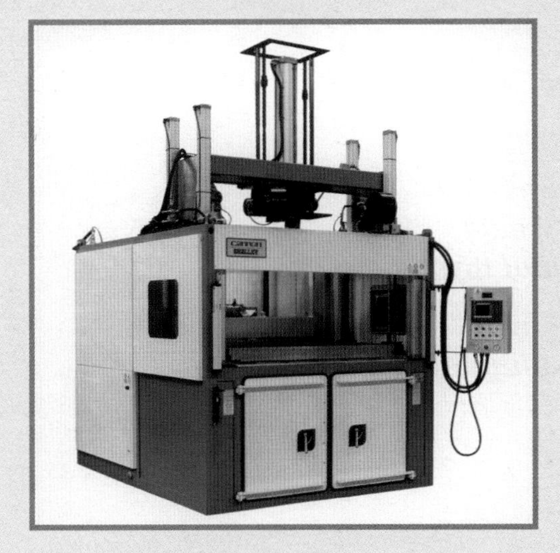

An industrial vacuum former.

7. This question is about manufacturing. See pages 40-41.
(Total 4 marks)

White marks appeared on the edges of the finished vacuum forming when you took out the former.

a) Write down the reason for these marks. *(2 marks)*

b) Write down two ways to stop this from happening. *(2 marks)*

8. This question is about researching. See pages 14-15.
(Total 2 marks)

A company that specialises in the manufacture of Christmas tree decorations has asked you to design three new designs for next year. These will be a snowflake, a star and an angel. They would like the decorations to be made from either wood or acrylic sheet, but they must have impressed, decorated or textured surfaces.

Name two ways of finding more information about the type, range and quality of products that the company produces. *(2 marks)*

9. This question is about designing. See pages 36-37.
(Total 6 marks)

Draw two alternative designs for one of the decorations. One design should be in wood and the other in acrylic. Use colour and drawing instruments where appropriate and label your designs. Each Christmas decoration must be no larger than 50 x 60 mm. *(6 marks)*

10. This question is about manufacturing. See pages 60-61.
(Total 10 marks)

The company decides to choose the acrylic design.

a) Name two ways of decorating the surface of acrylic. *(4 marks)*

b) Explain how to carry out one way of decorating the surface of acrylic using notes and drawings. *(6 marks)*

11. This question is about packaging. See pages 78-79.
(Total 4 marks)

a) The decoration will be packed in expanded polystyrene. Give one reason for this choice. *(2 marks)*

b) Give one reason why this is not a good choice for the environment. *(2 marks)*

Total marks = 60

Project Three: Introduction

'Once A Tree' are a chain of shops which only sell products made from wood. They would like to develop their range of traditional toys.

Can you come up with a design which will appeal to young children and use sustainable resources in its manufacture?

Construction (page 100)

Drawing in 3D (page 94)

Dear

I am writing to enquire whether you would be interested in producing a wooden toy for children aged 3-6 years for sale in our shops. It is essential that the toy is made from renewable timbers, in keeping with our policy of maintaining the balance of nature for this and future generations to enjoy.

We are particularly interested in the idea of pull-along wooden toys with moving parts, based on a vehicle.

To save on production costs it would be helpful if the toy was made up from a series of standardised component pieces.

Our customers are looking for safe, quality products with a traditional appeal which also reflect our concern with 'green issues'.

Would you please submit a working prototype, appropriate production drawings and short report which clarifies how it will meet our requirements for it to be a 'green' product.

Yours sincerely

Managing Director

Clarifying the brief

What exactly is the Managing Director of 'Once A Tree' looking for? Make a list of her requirements.

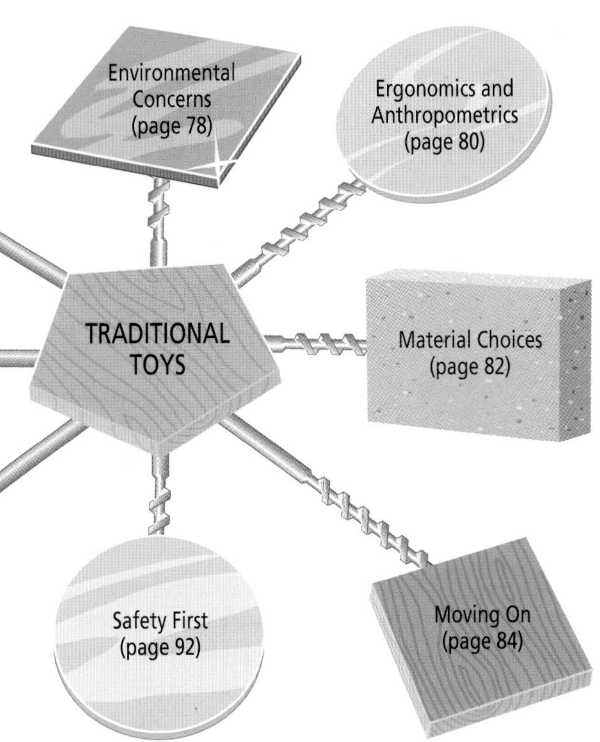

Get Started!

First the Managing Director wants ideas that she can discuss with you and choose from, not completed designs. You need to quickly come up with a range of ideas which you think she will like.

Sketch some possible vehicles. You could use photographs, posters, children's books or real vehicles as your resource. Make sure that they are stylized, that is that you take the shapes that each vehicle is made up of and simplify and exaggerate them.

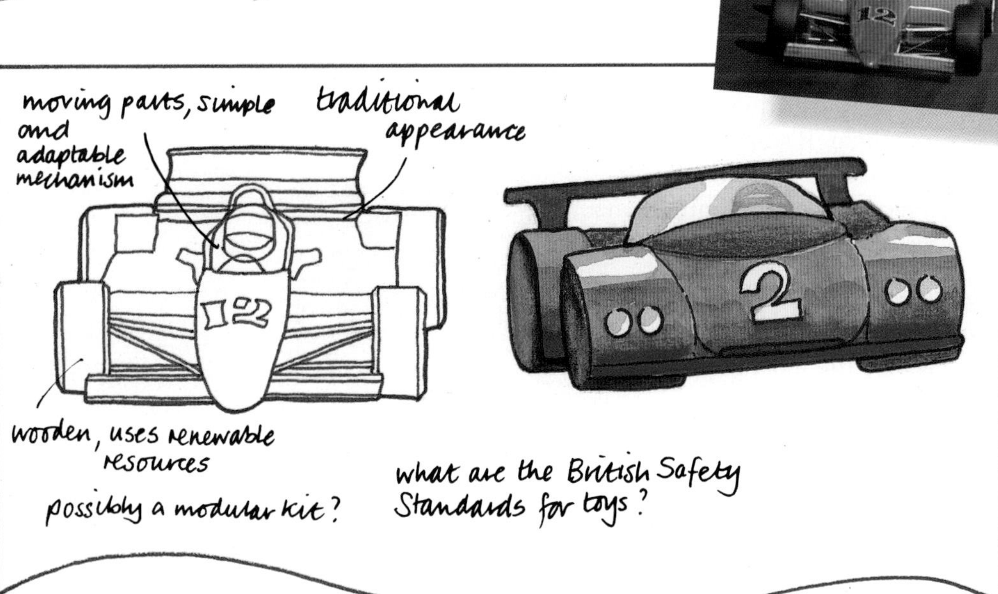

Get Moving!

Think about the parts of the vehicles which might move:

- a driver might bob up and down out of a racing car or train
- the container could tip backwards on a tipper truck
- the bucket on a JCB could be lifted, as could the cargo from the hold of a container ship.

Which mechanisms might you use for these movements? Could a single mechanism be used in more than one way?

Existing Solutions

You will need to find out more about the sort of toys young children like to help you with your designing. Evaluating existing toys will provide some useful information and also some ideas.

Toy Survey

Make a **survey** of children's wooden toys. You could write a **questionnaire** to help you to find out what the favourite wooden toys of a particular group of children are. As well as interviewing children, you should talk to adults, who are the people who usually buy the toys as presents.

You will need to think carefully about the questions which you ask and how you ask them. If someone doesn't understand a question, you will need to ask it in a different way. How will you record the information from the interview?

You could also present a group of children with a number of existing toys and carefully observe them playing with them. Which ones are the most popular?

When you have got all your results, how are you going to display them so that they are helpful to you and the Managing Director from Once A Tree?

- Who is the questionnaire aimed at?
- What am I trying to find out?
- Will the questions be understood?
- What sort of answers do I want – yes/no or more detailed?
- How will I be able to use this information?

Design Disassembly

Try to obtain an existing example of a child's toy which has mechanical moving parts. It is not necessary to take it to pieces. Examine it closely and use explanatory drawings to show what you think is happening inside.

- How do the components fit together?
- What are they made of?
- Which mechanisms are used?

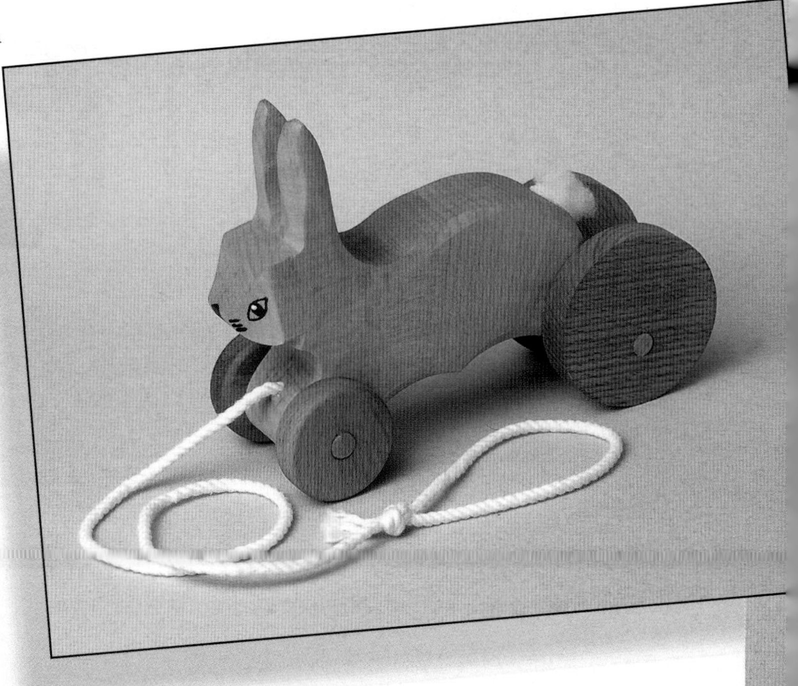

ICT 

Use a word processor to prepare your questionnaire, and a spreadsheet to show your findings graphically.

Toy story

All around the world, in every country and since time began, children have played with toys and games.

The ancient Egyptians played board games which used dice, such as backgammon and so did the Romans, while chess was adapted from an Indian war game played before AD 6.

Greek and Roman children played with hoops, tops and hobby horses.

In the 18th century dolls houses and model farms became popular. In the 19th century wooden soldiers, trains, musical boxes and kaleidoscopes became available.

Some toys are scaled down versions of the things which adults use in their everyday lives. Can you think of any examples? It is through playing with such toys that children learn about the adult world.

Other toys are used for comfort, for example dolls and soft toys. Greek children had clay dolls and Roman children had rag dolls made from cloth. Rich Victorian children had soft bodied dolls with painted china faces.

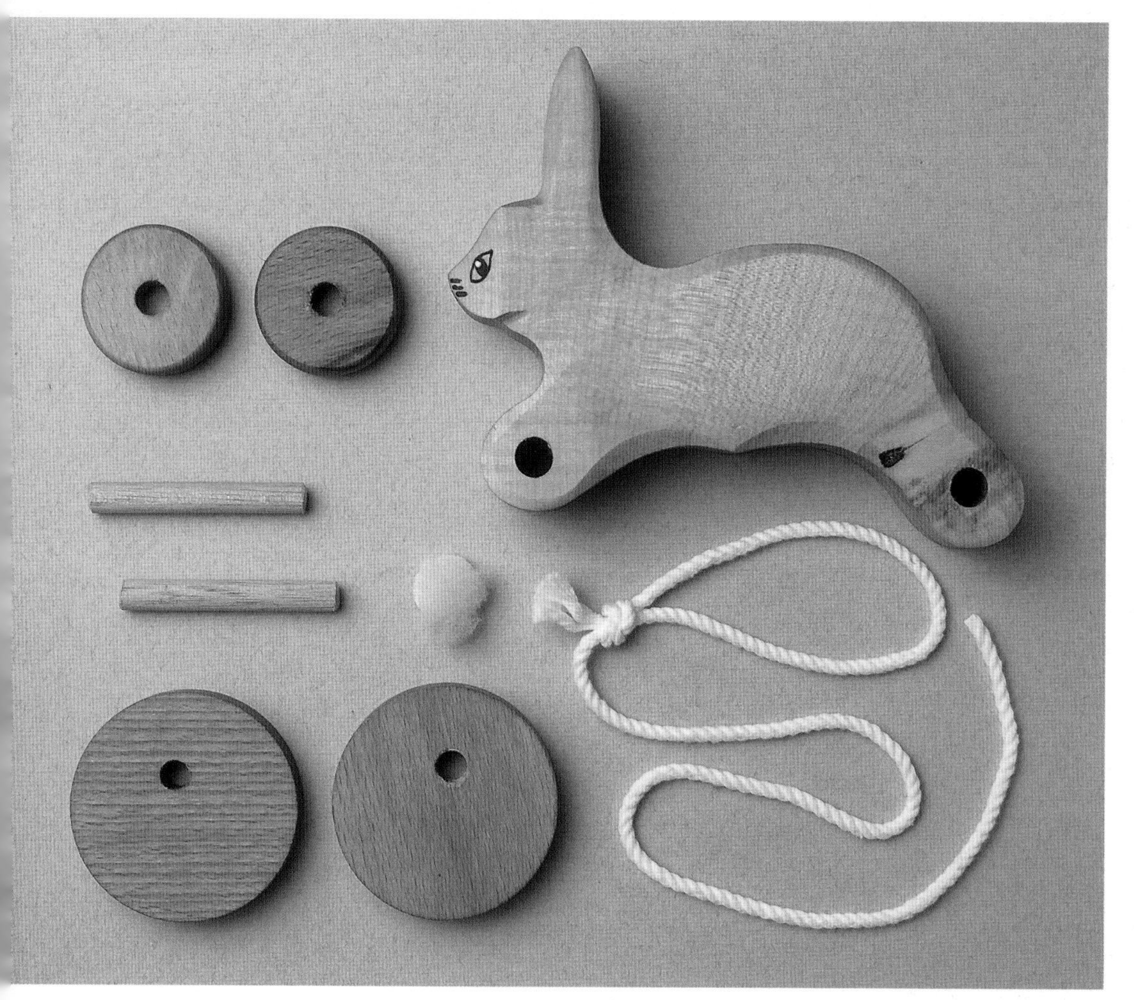

There are details of a range of mechanisms on pages 84 to 91 which will help you to work out how your toy works.

■ ACTIVITY

What specification do you think the designer of the toy you are analysing might have been working from?

See if you can work it out, using the following headings:

- use
- appearance
- materials
- finish
- safety
- durability
- ease of manufacture.

Environmental Concerns

Designers are responsible for ensuring that the products they create will minimise the potential damage to the environment. You will need to consider the different approaches to environmentally friendly manufacture.

One way of doing this is to remember to apply the 3Rs.

The 3Rs

Reduce

Aim to reduce the amount of:

- materials being used by making them thinner, lighter or smaller
- energy used in the manufacture of products
- packaging being used to protect products during transportation and storage.

Recycle

Consider if:

- the product could be made from **recycled materials**, e.g. glass, paper, aluminium foil, clothes, etc.
- it will be easy to recycle the product itself when it is no longer wanted.

Re-use

Try to specify parts and components (e.g. fastenings, packaging, etc.) which could be re-used easily when the product is finished with.

Did you know that...?

- Forests used to cover 66% of the world's land surface, today it is 20%.
- Forests are being cut down at a rate of 500 square kilometres a day – an area larger than Barbados.
- Everyday 6.5 million cubic metres of wood is harvested. 55% of this is hardwood and 45% is softwood.
- 50% of wood is used for cooking and heating. The developing world uses 90% of this.
- 50% of wood is used as timber, or wood products such as plywood and paper. Industrial nations use 90% of processed wood products.
- 3.5 million cubic metres is used for firewood.
- 1.2 million cubic metres is used for paper.
- 1.2 million cubic metres is used for sawn timber.
- 0.6 million cubic metres is used for other wood products.

Maintenance

Also think carefully about the **maintenance** of the product. Will it be easy to repair or replace individual components and parts when they fail, to avoid having to throw the whole product away?

An imaginative 3D pie chart can help show the data effectively

■ ACTIVITY

Create a graph or diagram which dramatically illustrates some of these figures.

Green Approaches

Less is more

One approach is to consider whether all the material being used for a product is really needed. The most creative design work often uses simple materials to achieve subtle and sophisticated products. These are also often cheaper to produce.

Reducing the amount of material used in surface area and thickness is an obvious approach, particularly where the product will not be subjected to a lot of wear and tear.

Litter

Reducing the number of parts in a package helps reduce the possibility of them being 'dropped' rather than thrown away. One familiar problem is a can ring-pull. How has this been solved?

Recycling

Glass is probably the easiest material to recycle and use again for the same purpose. Recycled papers and cards, rather than newly pulped and treated materials, are sometimes adequate for a job. Many types of plastics can be recycled but their chemical structure is weakened. As a result they may not be able to take up more complex shapes with precision.

Biodegradability

Biodegradable materials are those which decompose relatively quickly. Unfortunately this means that they also only give limited protection to the product. Biodestructable plastics is a technology still very much under development, but one day it might be possible to produce materials which decompose more rapidly than at present.

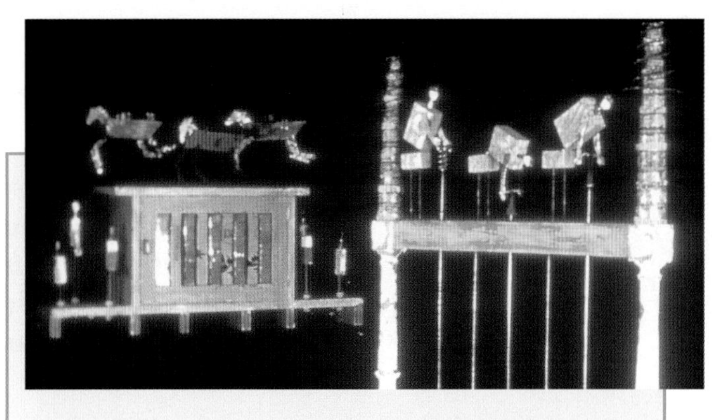

'I used to gather wood and tin from the beach, but now my materials mainly come from skips and Sunday morning market stalls'

Kristy Wyatt Smith

ACTIVITY

Read the various approaches to the problems of producing more environmentally friendly products above.

Find two examples of products which use one or more of these approaches. Draw them in colour on an A3 sheet, adding annotations to explain their design.

KEY POINTS

To make a product environmentally friendly:

- use the minimum amount of materials
- avoid materials which can't be re-cycled or re-used
- specify manufacturing processes which use the least energy.

Can-slab bench

'The cans are crushed together in a press and a new building material is produced. Its production requires minimal amounts of energy, causing little or no environmental harm.

The cans must first be sorted, washed and crushed into the blocks and shapes I want. Then I have to remove sharp edges by blasting and finally finishing and resin coating the pieces if required for extra strength and safety before final construction can take place.'

Jeremy Dent

IN YOUR PROJECT

When presenting an illustrated report to explain the ways in which your design will minimise the potential damage to the environment, remember to include details of:

- materials
- components
- packaging
- production processes
- paint and varnish finishes. See what you can find out about these from manufacturers' information leaflets.

Ergonomics and Anthropometrics

It is important to consider the needs of the user when designing products. The study of people in relationship to their environment is called ergonomics. The study of the human body and its movements is called anthropometrics.

Human beings come in all sorts of shapes and sizes. Males and females vary in their body proportions. We all have different likes and dislikes of colours, sounds and smells.

It is essential to identify the specific physical needs of your target customer when designing objects.

There are many things to consider, including the size, height, weight, strength and age of the people who will use the design and the range of movements they are able to make. For example, car designers need to consider the position of the car seat in relation to the control pedals, steering wheel and dashboard.

The designer of a child's toy has to consider how the needs of children may differ from those of adults. Size, weight and safety will be important aspects. Another is how the toy will appeal to the senses: young children learn through exploring the world using sight, sound, smell, touch and taste.

Ergonomics

Ergonomics is the scientific study of how people use buildings, furniture, tools, materials and equipment in their living and working environment. Designers use the results of these studies to help them design products which can be easily and effectively used. As well as telling them the sizes and shapes which will work well, they also need to know how people use their senses.

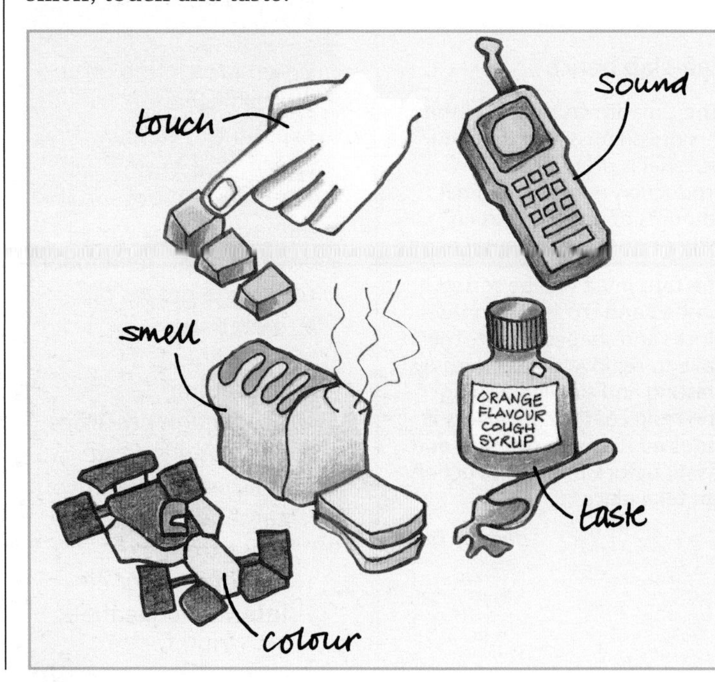

The senses

Think about the noises made by the everyday products which we have in our houses. As well as the telephone, microwave or door bell, what noise does the fridge door make when you close it? What sounds do you hear when you turn the light off or close the car door?

Now think about going round a supermarket. Some supermarkets have a range of fake 'smells' which are released at particular points in the store, for example a bread smell near the bakery.

People respond in different ways to different colours, textures, patterns, shapes and lines. These visual elements provide important clues as to how something works, and attract us to want to use the product.

Have you ever picked up a product and noticed how good it feels – the balance of a saucepan or a good fountain pen, for example? The feel of an object is very important for young children.

All these sensory qualities make a great deal of difference to the way in which people enjoy the experience of using a product.

Anthropometrics

Anthropometrics is the name given to the study of the average dimensions of human beings. By taking measurements from a large sample of people who will be using a particular product, the maximum, minimum and average sizes can be calculated. The design can then be made suitable for a wide range of people.

This data is used whenever products for human beings are designed, for example car seats and clothing. There are publications available which list this information, including for instance details of the range of human reach and ideal working surface height, both useful when you are designing kitchen furniture for example.

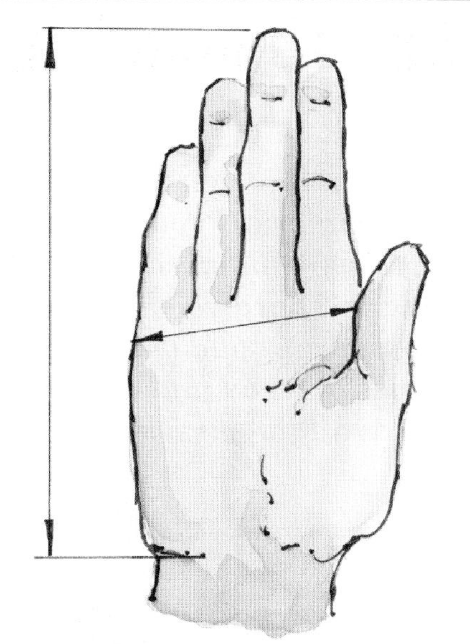

The hand length is measured from the crease of the wrist to the tip of the middle figure. The hand breadth is the measurement of the palm at its widest point, excluding the thumb.

ACTIVITY

Each person in the class should measure their hand length and breadth.

- What is the range of sizes?
- What applications can you think of for this information?

British children	Hand length (mm)	Hand breadth (mm)
3 to 4 years	Boys 100 –125 Girls 100 – 130	Boys 50 – 60 Girls 45 – 60
5 to 7 years	Boys 115 – 145 Girls 110 – 145	Boys 55 –70 Girls 50 – 65

If you were designing something to fit the hands of all children between the ages of 3 and 7, what variation in hand length and breadth would you need to take into account?

IN YOUR PROJECT

- What kinds of data would be useful to have?
- How could you find out about people's physical capabilities?
- How could you find out about what shapes, colours and textures, sounds and smells users will find attractive?

ACTIVITY

Whereas clothing for young children comes in a variety of sizes, toys need to be used by a range of different sizes of child.

You could arrange to visit a playgroup or nursery and measure the children, or ask everyone in your class to measure two children and share the data.

- Which dimensions will be useful to you? How will you measure them?
- How are you going to analyse and present the data in a useful way? Do you have access to a computer programme which will do this for you?

Environmental conditions

As well as designing for appeal to the senses, designers have to ensure that products do not have a harmful effect on the user. For example they need to consider irritating background noise, such as high pitched machinery noise.

Other problems could be caused by heat or cold, the positioning of an electric fan could cause a draught or the heat exhaust from machinery may lead to an overheated work environment. If you look around your workshop you will see examples of dust extraction systems over machinery and equipment which produce a lot of dust such as sanders and band saws. Other equipment, such as the hearth area or the brazing bench may have fume extraction. These extractors are installed for health and safety reasons, as well as to make the environment more pleasant to work in.

KEY POINTS

- Ergonomics is the study of people in relation to their living and working environment.
- Anthropometrics is the data which concerns the dimensions of human beings.
- Designers use ergonomics and anthropometrics to design products which people will find easy and satisfying to use.

Material Choices

It is important that you think carefully about the type of materials you use for your design. There are many hundreds of types of softwoods and hardwoods available. Each has its own properties and characteristics. Check which types of wood and manufactured board you have available in school.

IN YOUR PROJECT

You will need to think about the properties which the material you are going to use must have.

Softwoods and Hardwoods

There are two main categories of wood. **Softwood** comes from coniferous trees (those which keep their leaves in winter) and **hardwood** comes from deciduous trees (those which lose their leaves).

The names softwood and hardwood do not describe the qualities of the wood. For example balsa is a hardwood, yet it is very soft and easily worked or damaged. European redwood (Scots, Pine and Red Deal) is a softwood, yet it is hard and durable.

Softwoods

Name	Country of origin	Colour	Characteristics	Properties	Uses
European redwood (Scots, Pine, Red Deal)	British Isles, Scandinavia	Pale to reddish brown	Strong, hard, straight grain with some knots which tend to stay in	Works easily and durable when preserved. Can be painted or varnished	Furniture, window and door frames, floorboards, packing cases, toys
Sitka Spruce (whitewood)	British Isles, USA, Canada, Europe	Cream, golden yellow, brown	Resists splitting but has resin pockets and knots fall out	Tough and durable but easily worked. Can be painted or varnished	Constructional building work (rafters, floor and roof joists), paper making, packing cases and low cost furniture
Parana Pine	South America	Pale cream to brown and purple	Tough with a fine grain and few knots	Tends to shrink rapidly on drying and to twist and so is used in jointed constructions	Shelves, cupboards, fitted furniture, step ladders
Douglas Fir (Oregon Pine, Colombian Pine)	Canada, USA	Gold to red-brown	Strong, straight grain, resinous	Durable outside when used with a protective finish	Furniture, plywood, doors, window frames

Hardwoods

Name	Country of origin	Colour	Characteristics	Properties	Uses
Ash	British Isles	White to light brown	Long grain	Flexible, tough, can be steam bent or laminated	Sports equipment (cricket stumps, hockey sticks), furniture, toys, hammer shafts, shop and office interiors
Oak	British Isles, Europe, Japan, USA	Yellow, silver, pale brown (depending on species)	Heavy, strong, hard and durable	English oak is hard to work, Japanese oak is easier to work, but less strong	Furniture, veneers, panelling, doors, windows, roofs, gates, fencing
Mahogany	Central and South America, West Indies, West Africa	Pink to reddish brown	African (Gaboon, Sapele, Utile) quite hard and strong	American easier to work	Indoor work only, furniture, panelling, veneers, pattern making
Beech	Europe, British Isles	White to pale brown, speckled appearance	Hard and strong with a close grain, works easily but not durable outside	Good for turning on a lathe	Furniture, toys, tools, kitchen utensils, e.g. rolling pin, spatula, steak mallet
Balsa	Central and South America	Pale yellow	Light	Soft	Model making, rafts, life belts

Manufactured Boards

All **manufactured boards** are made from natural timber. They are either made from thin sheets (veneers) of wood which are glued together and compressed, or from particles of wood which are mixed with glue, then compressed and heated.

The advantage of manufactured boards is that they are stable, unlike timber which tends to absorb moisture and then dry out again according to the weather, twist and split. Also if you need to use a wide board, you are limited by the width of the tree, whereas a manufactured board can be produced to greater dimensions than it is possible to cut from a tree.

■ ACTIVITY

Look around the room and try to identify items which are made from manufactured boards, softwoods and hardwoods. Make a list of each. Which list is the longest? Why do you think this is? (Don't be fooled by appearance, remember that some cheaper materials are veneered or laminated.)

Hardboard

Made of very fine wood particles, mixed with glue, heated and compressed. It is very cheap but not very strong, and is used for the backs of kitchen cupboards.

Plywood

Made of an uneven number of layers of veneer (three or more). This stabilises the board as the grain of each layer runs in alternate directions. Decorative veneers are sometimes glued to the surface to give the impression of expensive timber. Birch ply is the best quality, Douglas Fir ply is used in the building industry for shuttering (the moulds into which concrete is poured) and WBP (water and boil proof) ply, which is made using waterproof glue, is suitable for use in water. It is relatively cheap and is used in a wide variety of applications including boat building, furniture and sports equipment.

Chipboard

Made from wood particles and glue, compressed and heated. It is often laminated with veneer or plastic laminate such as Formica which gives it strength. A relatively cheap material, it is used for kitchen and bedroom units and flat pack furniture.

Medium Density Fibre Board (MDF)

Made in the same way as chipboard and hardboard, but with very fine particles. It machines well, especially when routered. It is not waterproof and can be an irritant so a mask must always be worn when working with it. It takes stain and paint very well. Relatively expensive, it is mainly used for furniture.

Moving On (1)

***There are four different types of motion, and a variety of ways to change one type of motion to another. One way is to use a* cam*. Another is a* crank and slider mechanism.**

Motion

There are four basic types of **motion**.

Linear motion

This is where an object travels in a straight line, for example a ski lift or a train.

Reciprocating motion

This is where an object travels backwards and forwards in a straight line, for example a sewing machine needle, a piston in a car engine, a pogo stick and a pneumatic drill.

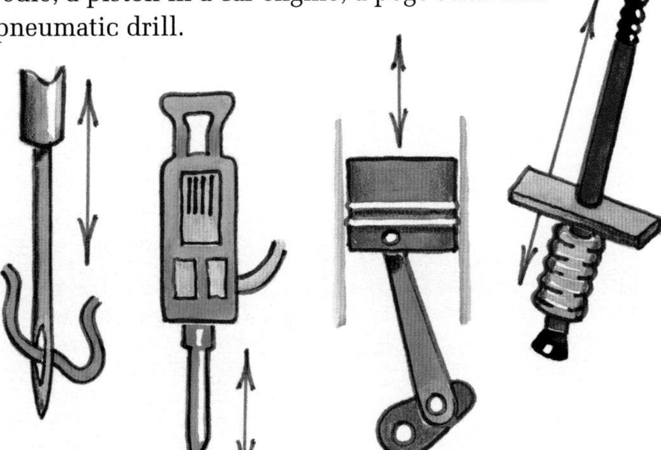

Rotary motion

This is a circular motion such as the blades on a helicopter, the big wheel at a fairground and the hands on a clock.

Oscillating motion

This is similar to reciprocating motion in so far as it is a backwards and forwards motion, but it follows a curved path rather than a straight line, for example a pendulum on a clock and a swing seat.

Which type of motion is used by each of these fairground rides?

Cam Toys

Look carefully at the drawing of the driver in the racing car and draw the **cam and follower mechanism** which makes the head bob up and down.

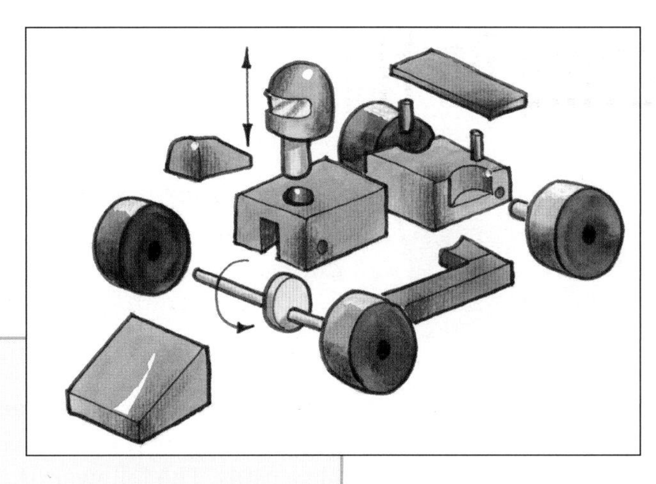

■ ACTIVITY

Carefully follow the instructions and the diagrams below and make a simple cam and follower toy for yourself.

1. Take four 80 mm lengths, and eight 90 mm lengths of 10 mm square section softwood.
2. Join the pieces of the 90 mm softwood as shown in the diagram to make two 100 mm square frames.

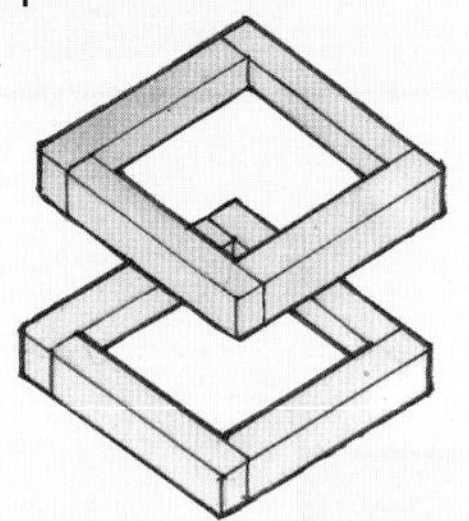

3. Make these into a 100 mm cube by joining the two frames together with the 80 mm lengths.

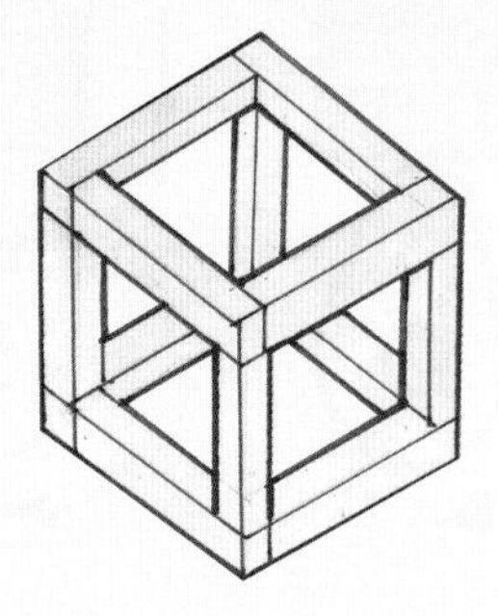

4. Now make a simple cam by drilling an off-centre hole into a disk of MDF or ply.

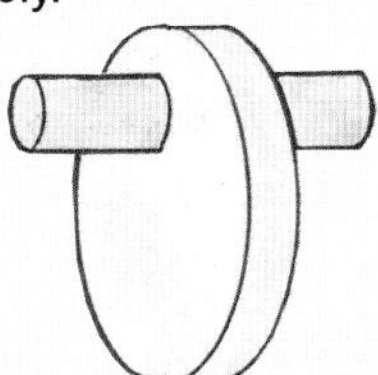

5. Push a 150 mm piece of dowel through the hole in the cam. Make sure that it is a tight fit so that it doesn't revolve on the dowel. You can glue it if it is not tight enough.
6. Hold your cam in the 'up' position against the side of your cube and calculate how far down the cube you will need to fix your dowel so the cam stays 30 mm below the top of the cube.

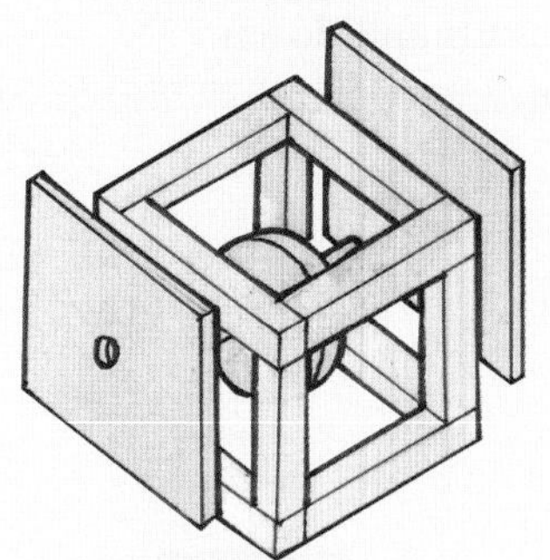

7. Punch holes in two 100 mm square pieces of card at this position and place the ends of the dowel through the holes in the card, one each side of the cam. Then glue the squares onto opposite sides of the cube.
8. Turn the dowel and your cam should turn too, supported on the dowel. You could attach another circle of wood or card to the dowel as shown to make a crank handle which makes it easier to turn.

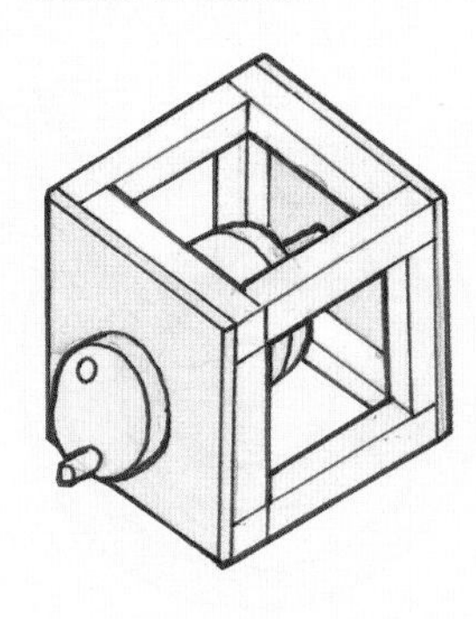

9. Make a follower for your cam by fixing a piece of dowel through a piece of card or wood and resting this on top of your cam. When the cam rises and falls, so will the follower.
10. Experiment until you get the distances right.

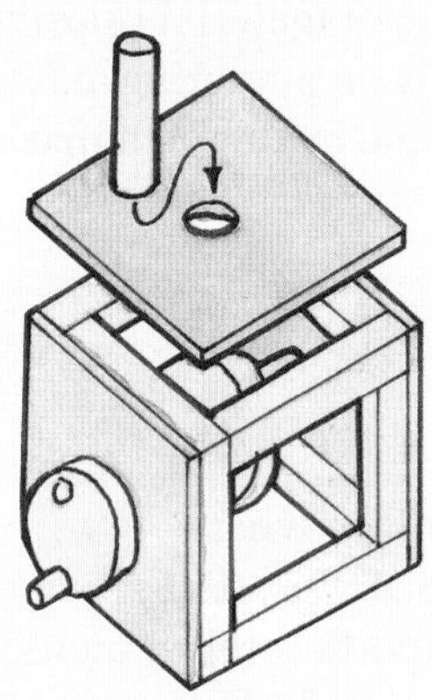

11. You can support the follower by placing another square of card over the top of the cube and making a hole in it which is big enough for the dowel to move up and down in.
12. Experiment until you get the distances right.

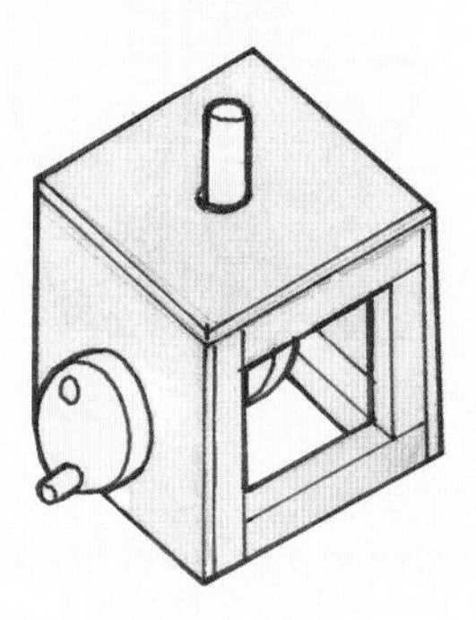

What is the effect of not placing the follower exactly over the cam?

What could you attach to the dowel to make a simple toy?

Produce a sheet of initial ideas for your moving toy design. Include the details of how you would include a mechanism to get part of a toy to move.

ICT 

Use a CAD program to simulate the movement of a mechanism.

IN YOUR PROJECT

- Think about the kinds of mechanisms which you might be able to use to make the parts move in your own designs.
- Don't forget to refer to the details of a range of mechanisms on pages 86 to 91. You don't have to use a cam and follower.

Moving On (2)

The two most useful cams for you to use in a design are likely to be linear cams and rotary cams

Linear Cams

Linear cams are used to change the direction of reciprocating motion. They are also known as flat plate cams. The follower usually has a wheel or roller on the end and rests on the cam. As the cam moves in a reciprocating motion, backwards and forwards, so the follower moves up and down, again in a reciprocating motion.

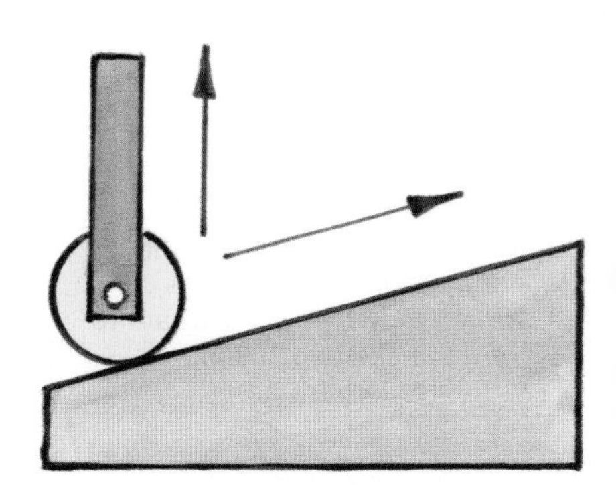

Rotary Cams

Rotary cams change rotary motion into reciprocating motion. The cam is a shaped piece of material (plastic or metal if inside an engine) which is on a rotating axle. The follower rests on the edge of the cam which makes the follower go up and down as it turns.

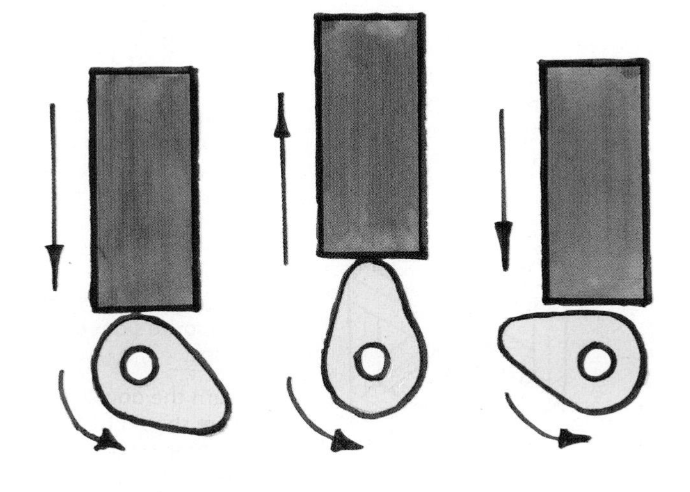

Rotary cams are used inside machines where reciprocating movement takes place. For example in car engines they play a number of important roles, including opening and closing valves, operating fuel pumps and opening and closing contact breaker points.

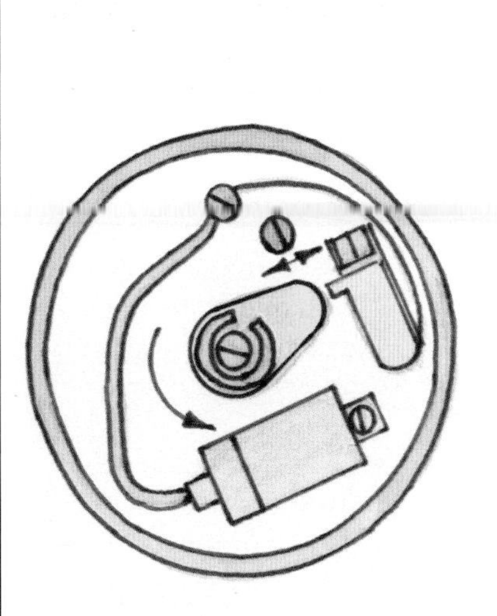

Contact breaker

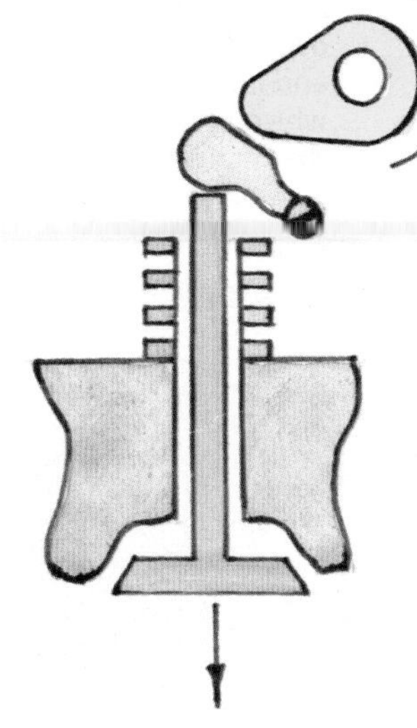

Valve in a cylinder head

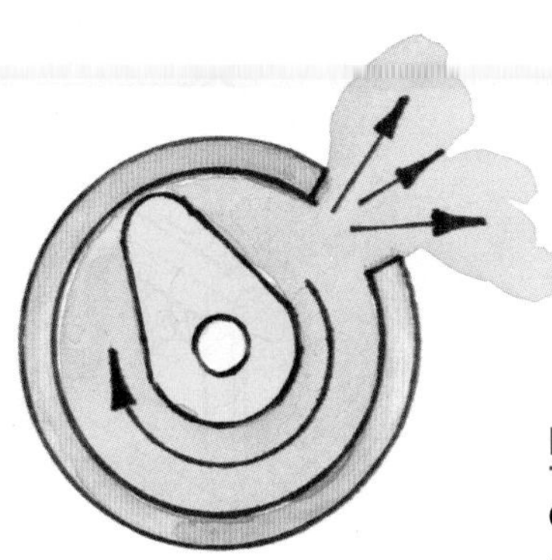

Fuel pump

■ **ACTIVITY**

Can you find out and explain how these work? Use diagrams, and label the different parts.

Cranks

A crank works on the same principle as a wheel and axle. The distance between the axle and the handle increases the leverage, which in turn allows more force to be used to turn the axle.

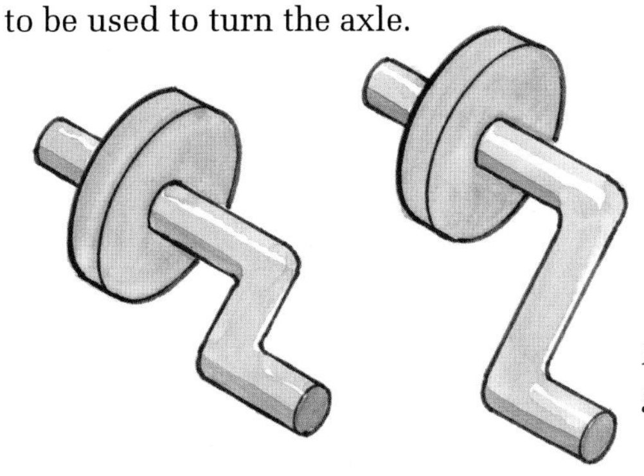

Increased distance gives increased leverage

Joining several crank handles together forms a **crankshaft**. This is the mechanism found in a child's pedal car and tricycle.

Crank and Slider

A **crank and slider** can be used to change reciprocating motion to rotary motion or vice versa. A crankshaft is joined to a piston via a connecting rod which is fixed onto the crankshaft at one end and the piston at the other. Both must be moving, as opposed to fixed, joints.

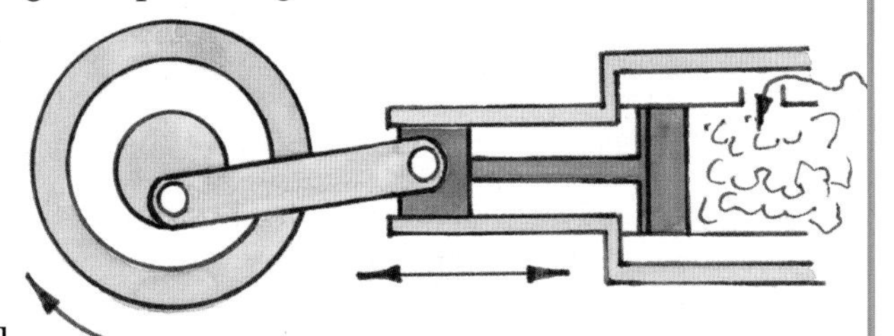

In a steam train, the engine is attached to the wheels by means of the connecting rod. As the piston is pushed in and out so the connecting rod makes the wheels go round.

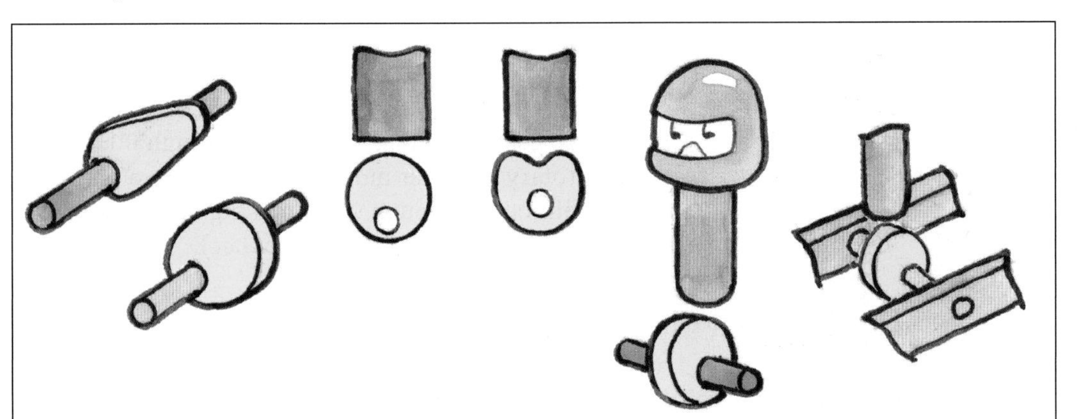

ACTIVITY

A power hacksaw uses the same principle, but the input is from the motor to the crankshaft.
Draw the crank and slider mechanism of a powered hacksaw.

KEY POINTS

- Cams are used to change one type of motion into another.
- Two common types of cam are the rotary cam and the linear cam.
- The rotary cam is like an odd shaped wheel which rotates on an axle.
- The linear cam is a slope which gives a reciprocating motion.
- The crank handle is like an axle with a 90° bend in it.
- The longer the distance between the axle and the handle, the better the leverage.
- Several crank handles joined together form a crankshaft.
- A crank and slider can change reciprocating motion to rotary motion, and vice versa.

IN YOUR PROJECT

How might you use:

- a rotary cam?
- an eccentric cam?
- cams to change rotary to reciprocating motion?
- a linear cam?
- a crank and slider?

Moving On (3)

Gears are used to transmit motion. They can also speed motion up, slow it down, and change its direction.

Getting Into Gear

Gears are toothed wheels which mesh with each other so that when one gear is moved, so do the others which are meshed with it. This is a way of transmitting rotary motion. The driver (the gear which is moved by hand or by using a motor) will rotate in the opposite direction to the driven (the gear which is meshed with the driver).

If you want them to rotate in the same direction, you need to put an idler gear between them.

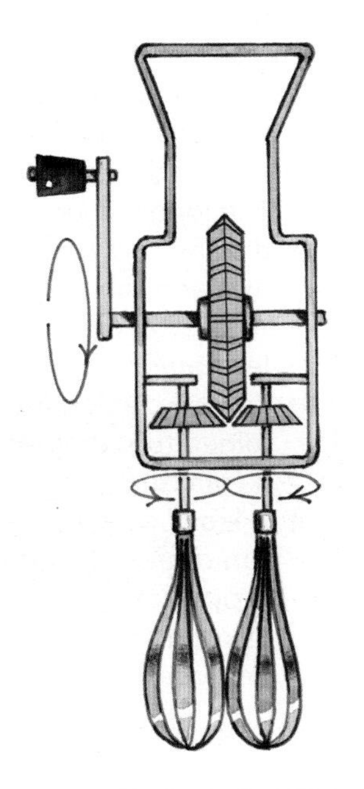

Speeding it Up and Slowing it Down

Gear wheels can be used to change speed. If you use two gear wheels of the same size there will be no change in speed. However, if the driven is smaller than the driver then there will be an increase in speed of rotation of the driven. If the driven is larger than the driver then there will be a decrease in speed of rotation of the driven. The larger the gear wheel, the more teeth it has.

Meshing a driver gear wheel with forty teeth with a driven gear wheel with eight teeth will cause the driven gear wheel to rotate five times for every single turn of the driver. This is called the **gear ratio** and is written 1:5.

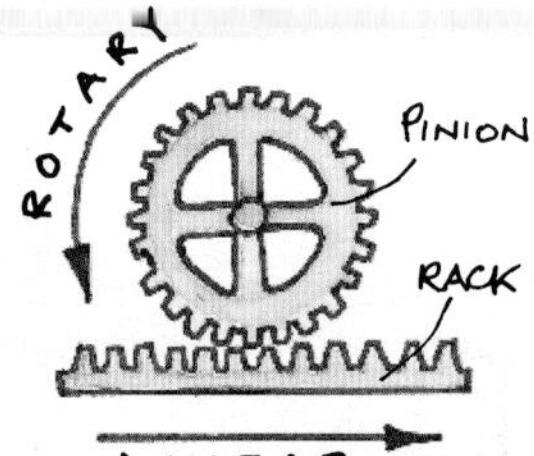

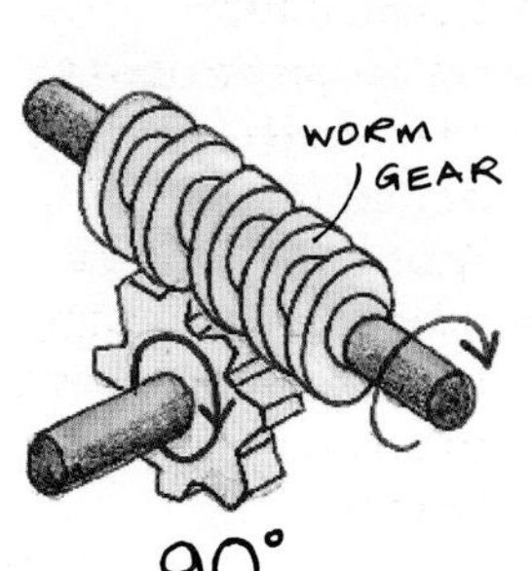

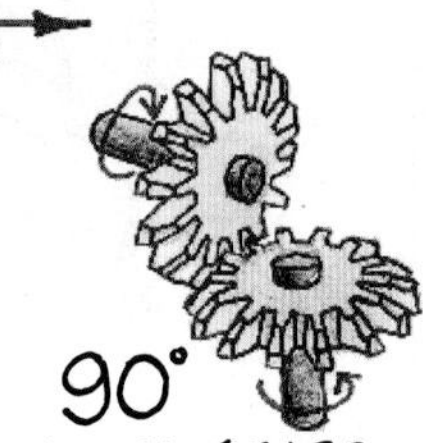

Changing direction

You can use gears to change direction through 90° and also from rotary to linear motion. There are two ways of changing direction through 90°, by using **bevel gears** and by using a **worm gear** with a worm wheel.

Chain and Sprocket

Meshed gear trains can be unreliable as the gears have to mesh perfectly or they will not turn smoothly. By using a **chain and sprocket system**, where the teeth on the gears do not have to mesh but are connected by a chain, you will avoid some of the problems. This only works for linear gear trains, not for bevel gears, worm gears or rack and pinion systems.

Speed ratio

The speed ratio between the driven and the driver may be calculated very easily:

$$\frac{\text{Diameter of Driver}}{\text{Diameter of Driven}} = \text{Speed Ratio}$$

In this example:

$$\frac{\text{Diameter of Driver}}{\text{Diameter of Driven}} = \frac{200\,\text{mm}}{50\,\text{mm}} = 4{:}1$$

Getting in Line

To change from rotary to linear motion, you use a **rack and pinion**. The gear wheel is the pinion and a straight toothed bar is the rack. If the rack is fixed, then the pinion will rotate and move in a linear direction along the rack.

If the pinion is rotating in a fixed position, then the rack will move in a linear direction along the pinion.

IN YOUR PROJECT

What types of movement might you be able to produce by using gear trains?

KEY POINTS

- Gears can be used to transmit motion.
- Gears can be used to speed up or slow down motion.
- Gears can be used to change the direction of motion.
- A gear train can be joined using a chain, this is referred to as a chain and sprocket system.

Moving On [4]

Here are some more ways of producing movement.

Levers and Linkages

Levers consist of a beam, a load, an effort and a fulcrum (or pivot). There are three classes of lever and the relative position of the **load**, **effort** and **fulcrum** will determine which class the lever belongs to.

A **first class lever** has the effort at one end of the beam, the load at the other and the fulcrum in the middle, for example a child's see saw, a pair of scissors (two first class levers joined together) or using a spoon to open a tin of treacle!

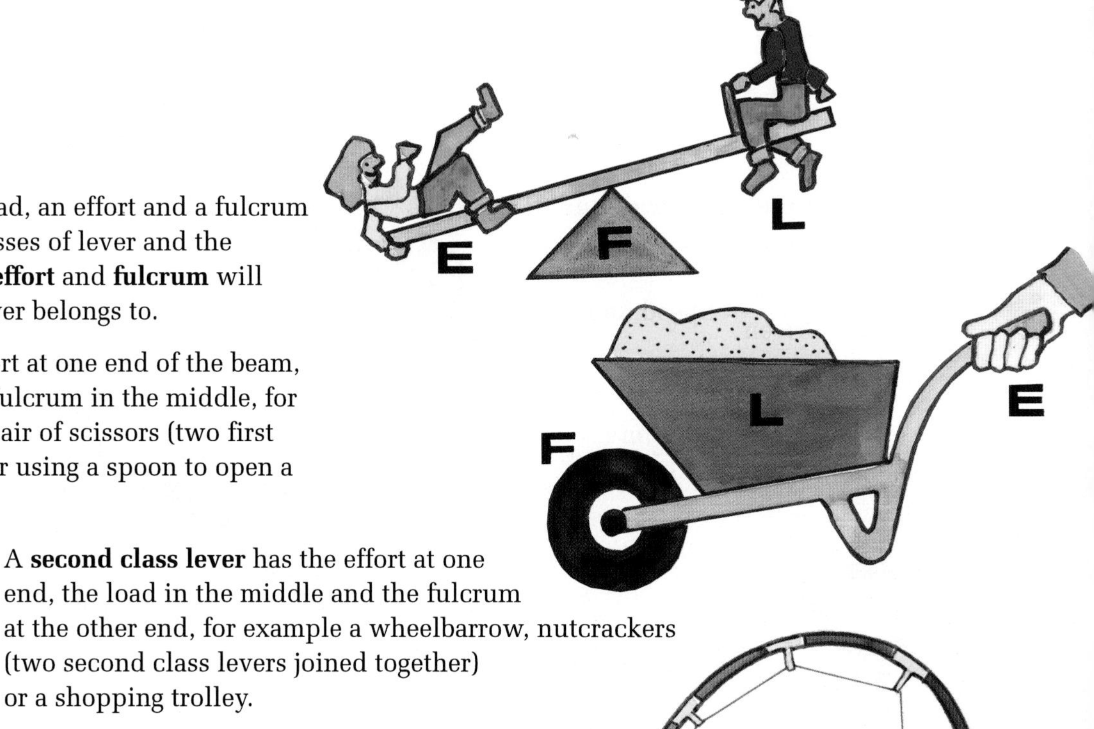

A **second class lever** has the effort at one end, the load in the middle and the fulcrum at the other end, for example a wheelbarrow, nutcrackers (two second class levers joined together) or a shopping trolley.

A **third class lever** has the effort in the middle, the load at one end and the fulcrum at the other end, for example a Suffolk door latch, a pair of tweezers (two third class levers joined together) and a fishing rod being used to land a fish.

All together now

When two or more levers are joined together, this is known as a **linkage**. There are many possible combinations of levers.

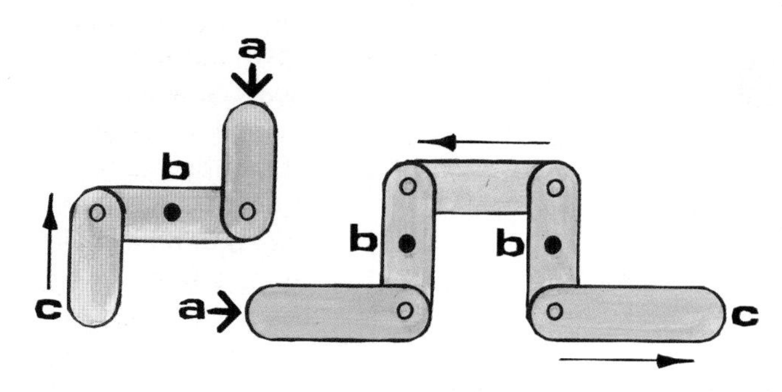

Just a moment

When a force (effort) acts on a lever it makes it rotate. This turning force is called a moment. Moments can be calculated to find out how much effort will be needed to move a load. The ratio between the load to be moved and the effort needed is called **mechanical advantage**.

$$\text{Mechanical advantage} = \frac{\text{load}}{\text{effort}}$$

$$\text{e.g.} \quad \frac{500\,\text{N}}{100\,\text{N}} = \frac{5}{1} = 5{:}1 \text{ or } 5$$

The larger the number, the greater the mechanical advantage. Class 1 and 2 levers give the most advantage. This means that a large load can be moved using a small effort. Class 3 levers are less common because their mechanical advantage is less than 1.

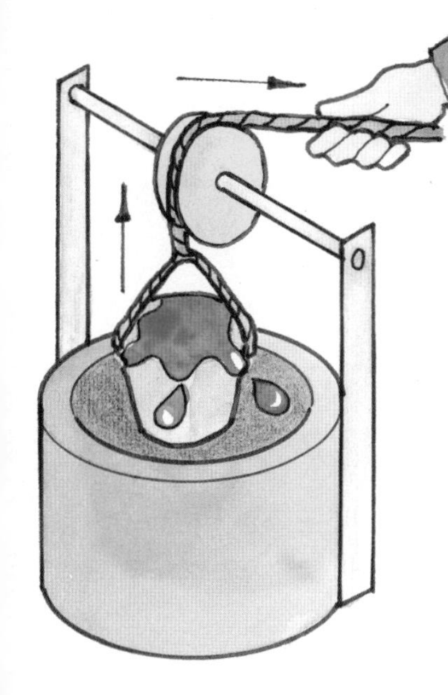

Pulleys

Just like levers, **pulleys** can make heavy loads easier to lift by using **mechanical advantage**. Pulleys can vary the mechanical advantage and can move a load when you exert a force in the opposite direction to that in which you want the load to move.

A fixed pulley

This type of pulley cannot produce mechanical advantage since the force required to move the load is the same as the load. They only allow a change of direction: you pull down on the rope to lift the load up.

A block and tackle

This consists of a fixed pulley and a moving pulley. It is capable of mechanical advantage since the load raised may be greater than the effort.

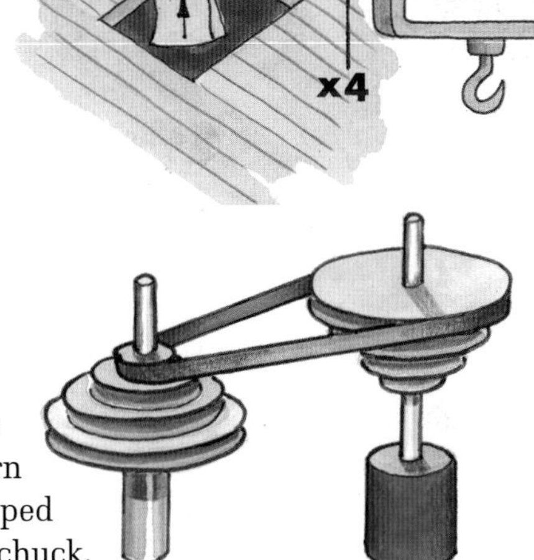

IN YOUR PROJECT

- How could you use levers or a linkage to make a moving part in your design?
- Also think about ways in which you might incorporate a pulley.

Belt drives

Belt drives work in the same way as chain and sprocket systems, but they need more friction to run, otherwise the belt will simply slip on the pulley. Belts will slip if the load is too great. This can be an advantage if a motor is driving a pulley joined to another by belt drive. Why do you think this is?

An example of belt drive is found in a pillar drill, where the driver stepped cone pulley is attached to the motor. This in turn is attached by belt drive to the driven stepped cone pulley which is attached to the drill chuck. In this way, by choosing various combinations of driver and driven, the speed of the drill can be varied.

■ ACTIVITY

Find out how well your ideas work by using strips of card with holes punched in them joined together with paper fasteners to make the linkages. Pin them with drawing pins to corrugated card or fibre board at the point marked b in each example, and apply a force at point a.

Try making up some linkages of your own and see what the output is (that is, what happens to the other levers in the linkage).

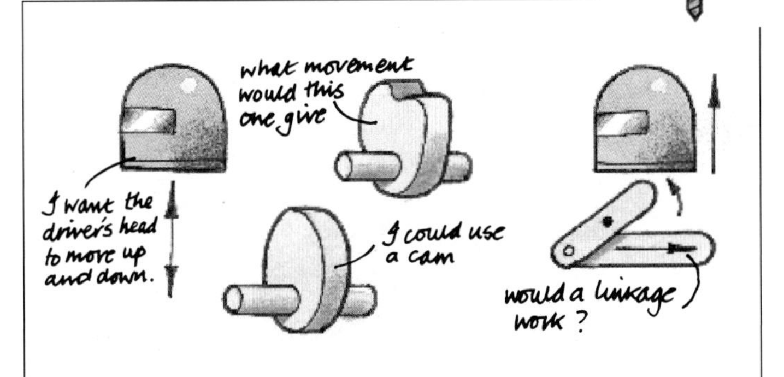

KEY POINTS

- There are four basic types of motion: rotary, reciprocating, linear and oscillating.
- Cams can change the direction of movement (e.g. linear to rotary).
- Levers can be joined together to form linkages.
- Pulleys and gears can transmit motion.

Safety First

Safety is one of the most important considerations when designing for children. It covers a range of different aspects including materials, finishes, size of components, finger traps and choking hazards, stability, durability and quality of manufacture.

CHECKPOINTS

Pull-along toys

✔ How thick is the cord on the toy? For children under 3 the cord should be more than 1.5 mm thick. This helps stop a child's fingers being cut if the toy gets stuck.

Toy cars and trains

✔ These must not have any sharp metal edges, tags or points. Edges should be rolled or folded over, or covered so that they cannot cut anyone.

✔ Wheels and tyres must be securely fastened.

Mechanical toys

✔ Use a pencil or pen to poke into any hole, gap or opening. If you can reach gears or other sharp or moving parts then the toy may be dangerous.

Consumer Toy Watch

Each year thousands of people are given hospital treatment for injuries involving toys. Most of these are caused by:

- falling over toys left lying around
- toys being misused
- broken toys.

But some are caused by toys which were badly made or designed.

The **safety laws** on toys in Europe are demanding. All toys need to meet the requirements of **British Standard BS 5665** and **European Standard EN71**.

The standard is in three parts:

- Physical and mechanical – all those parts of a toy that you can handle or touch. The standard tries to make sure that toys can't cut, stab, mangle or choke a child.
- Flammability – is the toy easy to set on fire if it is put near a flame, fire or heater?
- Migration of elements – this means anything which would be poisonous if swallowed or chewed.

CONFORMS TO
BRITISH STANDARD 5665

When a toy manufacturer makes a product which conforms to EC standards, the CE mark *should be put on the toy.*

■ ACTIVITY

Think carefully about the safety aspects of the toys shown on the left.

Which of the toys would be suitable for a two year old and which would be suitable for a seven year old?

For each toy which you do not think would be suitable for a two year old, give your reason.

Materials

Materials are chosen for their physical and aesthetic properties and when designing toys for children the choice is usually either wood or plastic.

In Victorian times and through the early years of this century, lead was used to make children's toys, especially model soldiers, resulting in many children suffering from lead poisoning. Lead also used to be a basic ingredient in paint manufacture, again posing a health risk until the hazard was recognised and a substitute for lead was found.

Although plastic comes in a range of bright colours and is cheap and durable, it can be brittle and crack easily, leaving sharp edges which can easily cut fingers and mouths. Small pieces can be swallowed and this poses a particular problem with rattles which are often also filled with small pieces of plastic to give the rattling effect when shaken.

Wood is a traditional material used for toys, but it can splinter easily. Manufactured boards, such as plywood, can deteriorate if left outside, leading to potentially dangerous jagged edges. Other woods have an irritating effect on the skin. Some manufactured boards are made using adhesives which could pose a health problem if placed in the mouth. So choose carefully.

Finishes

Applied finishes need to be non-toxic since children often put things in their mouths. They should be resistant to scraping and flaking, since small children usually handle toys very roughly. The finish should be weather-proof if toys are to be used outside. You can check on the toxicity and durability of finishes by referring to information on the container or on the manufacturer's information leaflets.

Finger traps and choking hazards

Finger traps are gaps or openings in a toy where children could injure their fingers and all such openings need to be large enough not to present a hazard.

To avoid being a choking hazard, individual components need to be large enough for children not to swallow them or to put them in their ears or up their noses. All components not designed to be removed need to be securely fixed.

Stability

Stability is important for toys on which children can ride. These need to have a low centre of gravity so that the toy does not topple over if the child leans over to one side. Swings and see-saws need to be firmly fixed to the ground.

IN YOUR PROJECT

- What are the particular safety requirements for your design?
- Who could you ask for more information?

KEY POINTS

- Safety is an essential consideration when designing toys.
- Consider materials, finishes, size of components, finger traps, choking hazards, stability, durability and quality of manufacture.

To find out more about safety standards go to: **www.bsi-global.com**

Toy dropped from 850 mm onto impact plate 5 times to see if drive mechanism is exposed by damage. BS5665

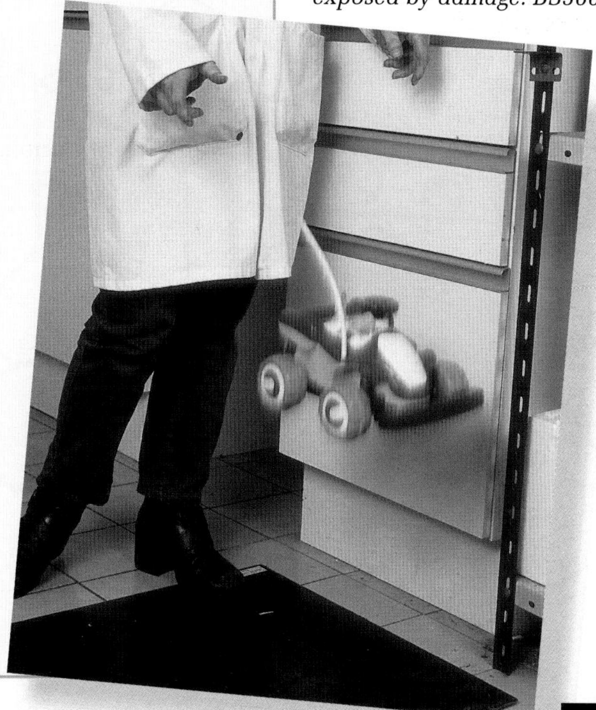

Drawing in Three Dimensions (1)

Three dimensional drawings range from a thumbnail sketch on a scrap of paper to a carefully constructed fully dimensioned image on a drawing board or computer screen.

Designers use 3D drawings to get a better idea of what the product they are designing will look like in reality, and to communicate their proposals to a client.

Freehand Sketching

Designers need to learn to sketch their ideas quickly in freehand. This may be for their own benefit as part of the process of visual thinking and development, or it may be to show to others for discussion.

Such sketches should not be drawn with rulers or other drawing aids, as this would slow the process down, and accuracy is not needed at this stage. They are usually in pencil, ink or marker pen, but you should experiment with different graphic media to develop your own successful style.

The main thing is to sketch confidently and boldly, and not be afraid to put pen to paper, or to make mistakes! Never try to erase or cover up mistakes – just try again on another part of the page.

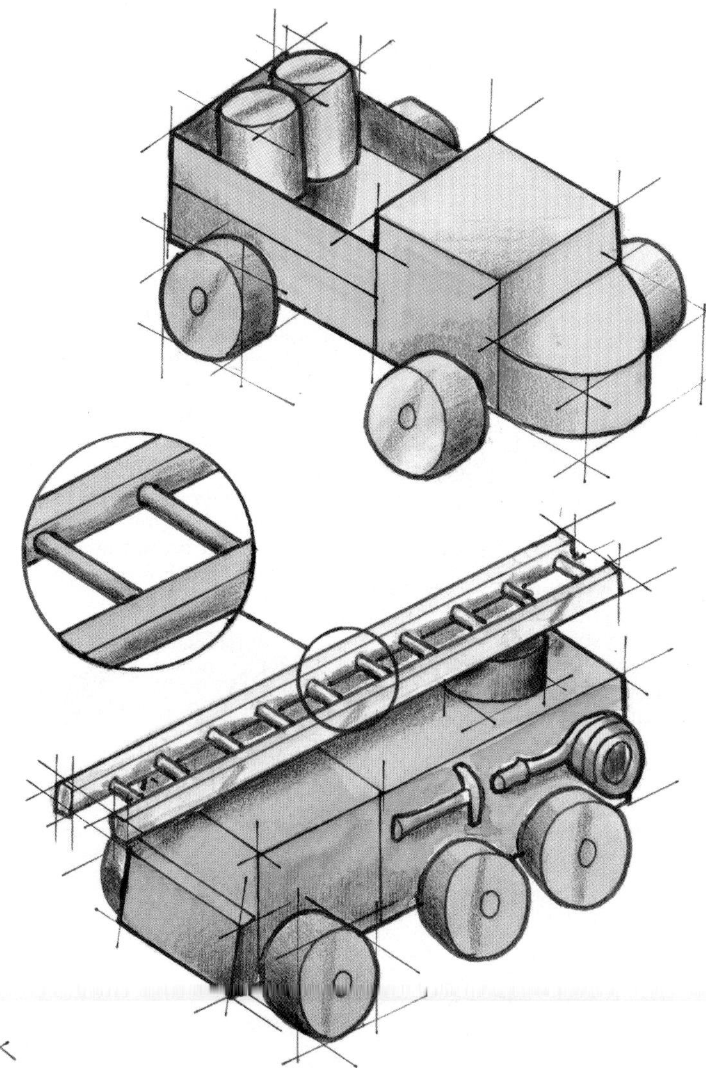

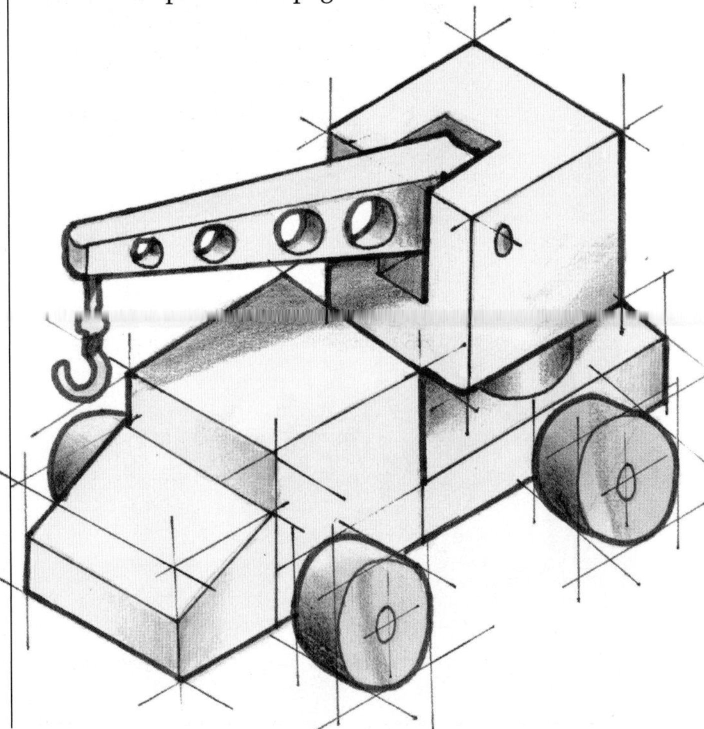

Crating

The method of 'boxing in' or '**crating**' is widely used to aid the construction of three dimensional drawings.

If you can master drawing cubes, rectangular boxes and cylinders in three dimensions it is surprising how many everyday objects, and of course your own design ideas, you will be able to draw convincingly.

Isometric Drawing

Isometric is a drawing system that is very useful and common in product design. A formal isometric drawing should follow certain guidelines, but it is also an excellent system for freehand sketching.

Isometric drawings are fairly realistic to look at, and are quick to do. They are particularly good for use in instructional drawings.

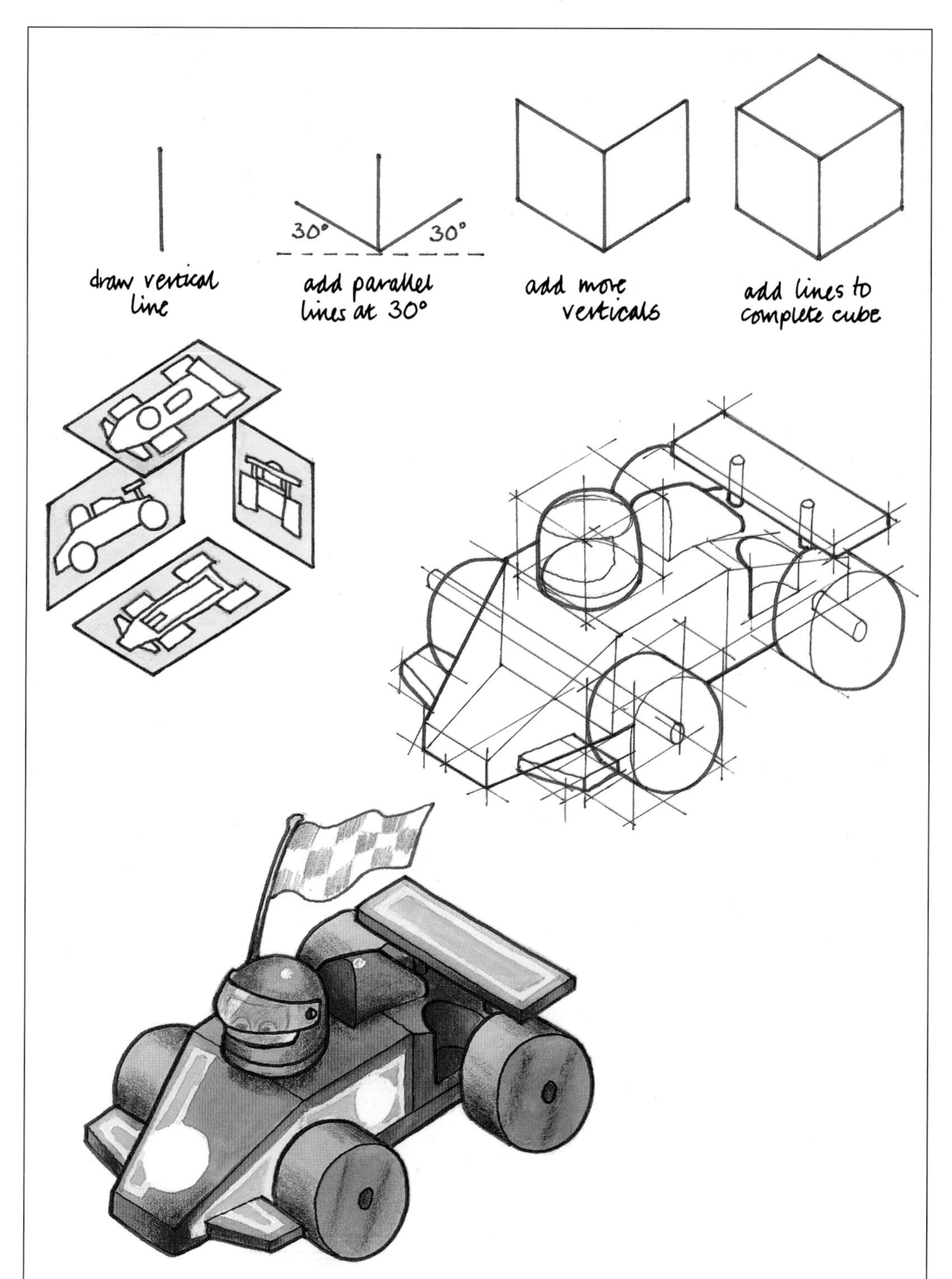

IN YOUR PROJECT

- Use freehand isometric sketches to help develop the design of three-dimensional objects.
- Use measured isometric drawings as the basis for the final presentation of ideas for products.
- Rendering techniques can be added to both sketch and formal isometric drawings.

ACTIVITY

- Obtain a child's toy. Choose one which is mainly rectangular in shape, with perhaps some curved forms. To begin with prepare an orthographic sketch of it (see page 38).
- Then attempt to develop an isometric drawing of its overall shape. Use the 'crating' approach to develop its other shapes and forms.
- Remove the construction lines and darken the lines you need. If you find the object too difficult to draw choose something simpler.
- If you find the object easy to draw in isometric, move on to something more demanding.

KEY POINTS

- Isometric drawing can be freehand or drawn using instruments.
- Freehand isometric is a useful system for quickly drawing ideas in 3D.
- Crating helps to establish the main forms of an object and its proportions.

Drawing in Three Dimensions [2]

A cut-away drawing is a two or three dimensional drawing with part of the drawing left out in order to show areas that would otherwise be hidden.

Exploded drawings are used by designers to show how things fit together and how they come apart. They also show clearly details that might otherwise be hidden in a drawing.

IN YOUR PROJECT

- Try to use simple cut-away and exploded drawings while sketching ideas.
- Also use them as part of your final presentation work to help illustrate key aspects of your design proposals.

Cut-away Drawings

When producing a **cut-away drawing**, the designer needs first to decide on the most suitable sections to leave out in order to show the detail required. Plans, elevations and isometric projections make a good starting point for isometrics.

The drawings themselves may be quite complicated to construct. Colour can be used to good effect to highlight certain details, but these types of drawings are often rendered in black line hatching to help identify different component parts.

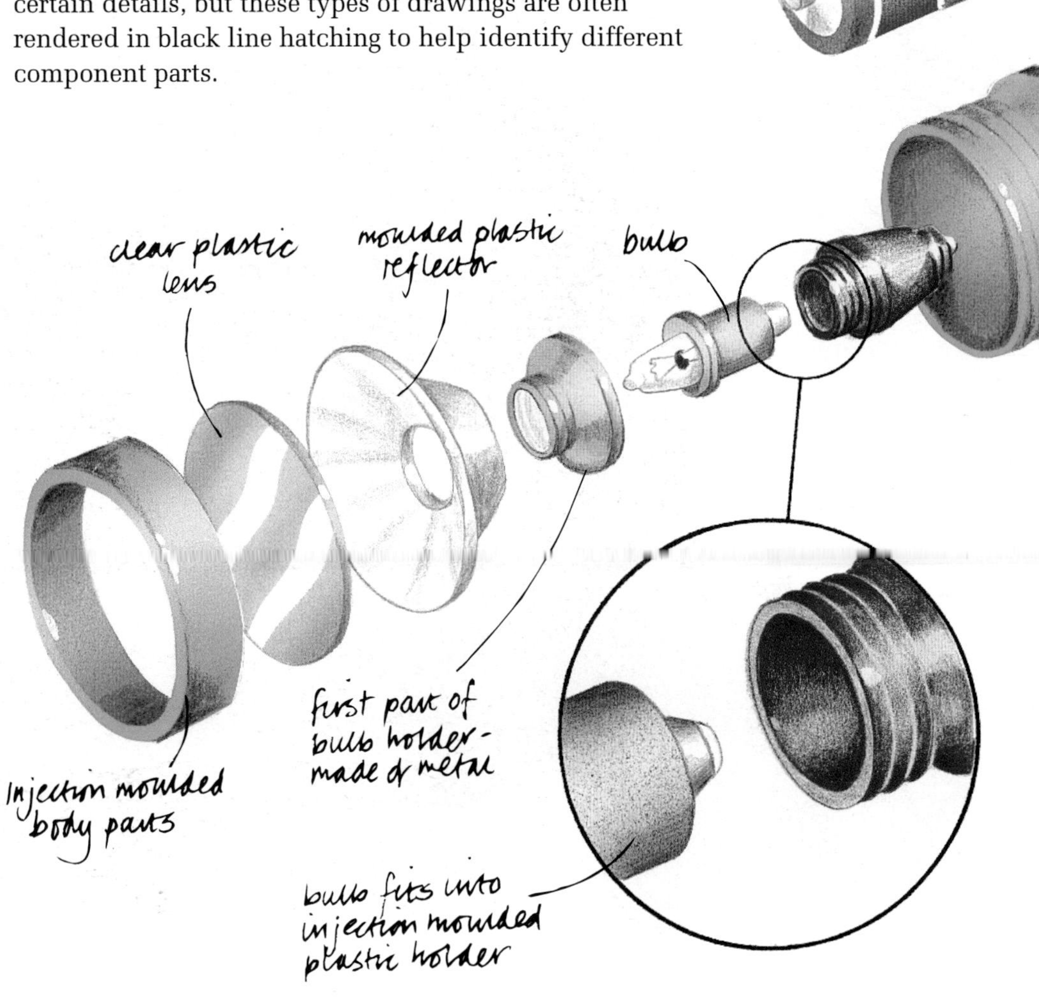

KEY POINTS

- Mechanical and electrical product manufacturers often use cut-away drawings to show the workings of the things they have designed. Architects and interior designers also use them to reveal what is behind wall surfaces.
- Exploded drawings are used by designers to show how things fit together and how they come apart. They also show clearly details that might otherwise be hidden in a drawing.

Exploded Drawings

Exploded drawings are often used in instruction or repair manuals. They help show clearly how a series of components fit and work together. An exploded drawing is usually shown in three-dimensions. Isometric projections are often used for this purpose.

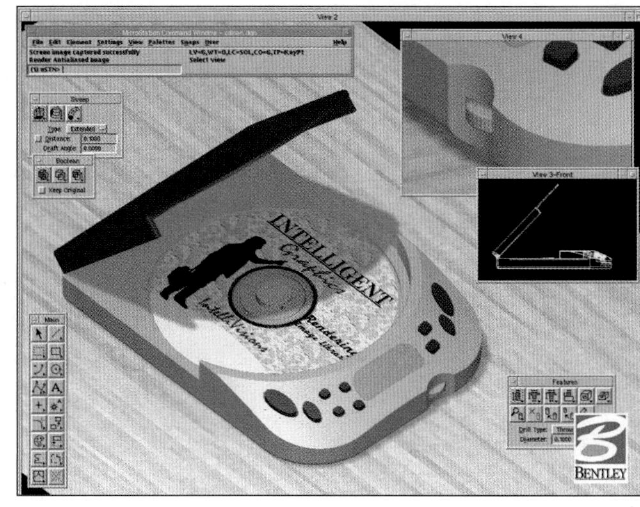

ICT

Coloured 3D drawings can be produced using CAD systems.

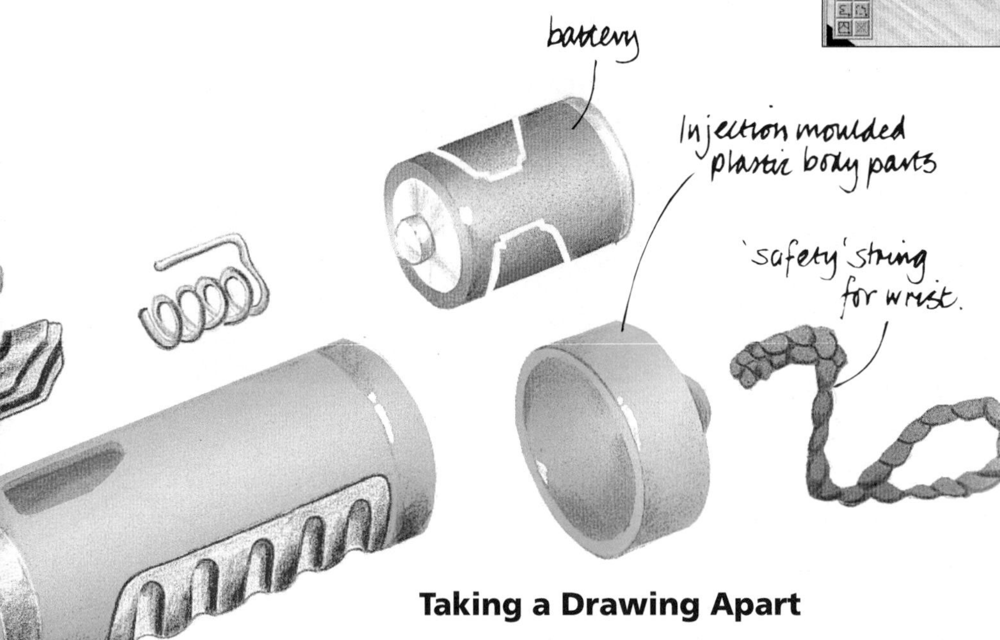

Taking a Drawing Apart

The individual parts of an exploded drawing are first drawn separately on tracing paper and moved away from each other, creating space in between the components. Careful visual positioning of the parts, using a light-box, helps makes them look like they fit together. The separated pieces are often shown overlapping the surface they have been extracted from. The result is a lively and highly informative drawing.

Some important things to remember are:

- exploded drawings are not usually dimensioned, although the various parts can be numbered or annotated.
- sometimes it can be effective to leave all or part of the drawing in freehand sketch form, rather than working up a highly technically controlled finish.
- colour can be added in an impressionistic or diagrammatic way to help communicate different components or materials.
- to experiment with different textured and coloured surfaces to represent the various materials which have been used.

ACTIVITY

See if you can obtain an old torch. Make sure you ask permission first, and that someone has checked it is safe – torches can get rusty, and acid from old batteries may have leaked out.

Carefully disassemble as much of it as you can, by unscrewing or pulling apart the various parts of the casing.

- What does each component do, and how does it do it?
- What is it made from?
- How does it fit with the other parts?

Think about how the torch has been assembled, and the materials and components which have been used. Identify some of its features which you think have been well designed and some which you feel are not very successful.

Present your findings by means of an annotated exploded drawing.

Four Wheels or Eight?

You need to start to develop your ideas about which vehicle you are going to choose to make and how it will be made. Which parts of the vehicle do you want to move and which mechanism will you use to achieve this? Gears are very difficult to make accurately. Will you make all the mechanism yourself, or include ready made components?

Developing Ideas

If you have decided upon a basic lorry or truck shape there are many different variations upon that theme. You could articulate it, that is separate the cab from the trailer at the back so that it will go around corners more easily.

You may want to add extra wheels to make it appear more robust.

If you have got as far as designing the cab and the chassis, what are you going to put on top? You could think about making a basic vehicle with exchangeable parts to make it into a number of different toys, for example a fire engine, a towing truck or a skip lorry.

How can you fix the components in place to stop them from falling off when the toy is moved? Can you think of a safe way of doing this?

is it safe to use?

The Mechanism

If you have designed the basic vehicle you will need to think about the mechanism. What will it need to do? Look back at the range of possible mechanisms on pages 84 to 91.

It is important that you fully understand the mechanism and how it will work before you start to make the vehicle, because you need to ensure that the mechanism is designed as part of the toy and that you have made allowances for it in your calculations of the dimensions.

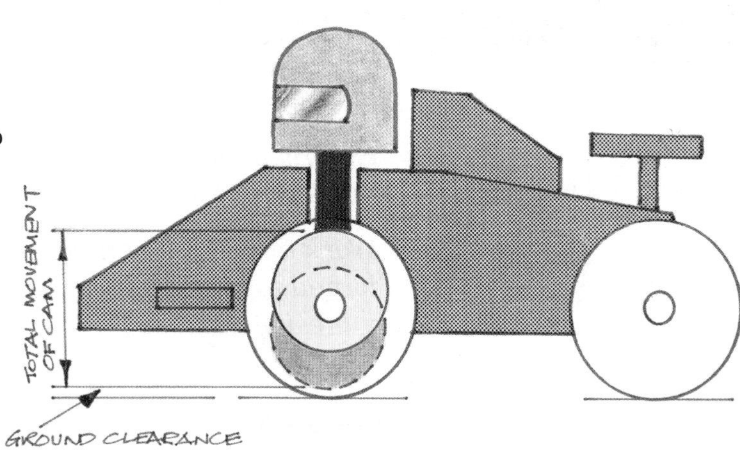

■ ACTIVITY

Model your chosen mechanism in 3D before you make a final decision, to make sure that it will do what you want it to. You could use the cube construction method on page 85 to hold all of the components of your mechanism in place while you work this out, or cut the walls out of a small cardboard box, leaving enough card around the edge of each face to support it.

Next Steps...

Once you have decided upon your final idea, you will need to do a working drawing. This should be an orthographic projection (see page 38), as this will help you to calculate the exact sizes of each part of the toy and the mechanism.

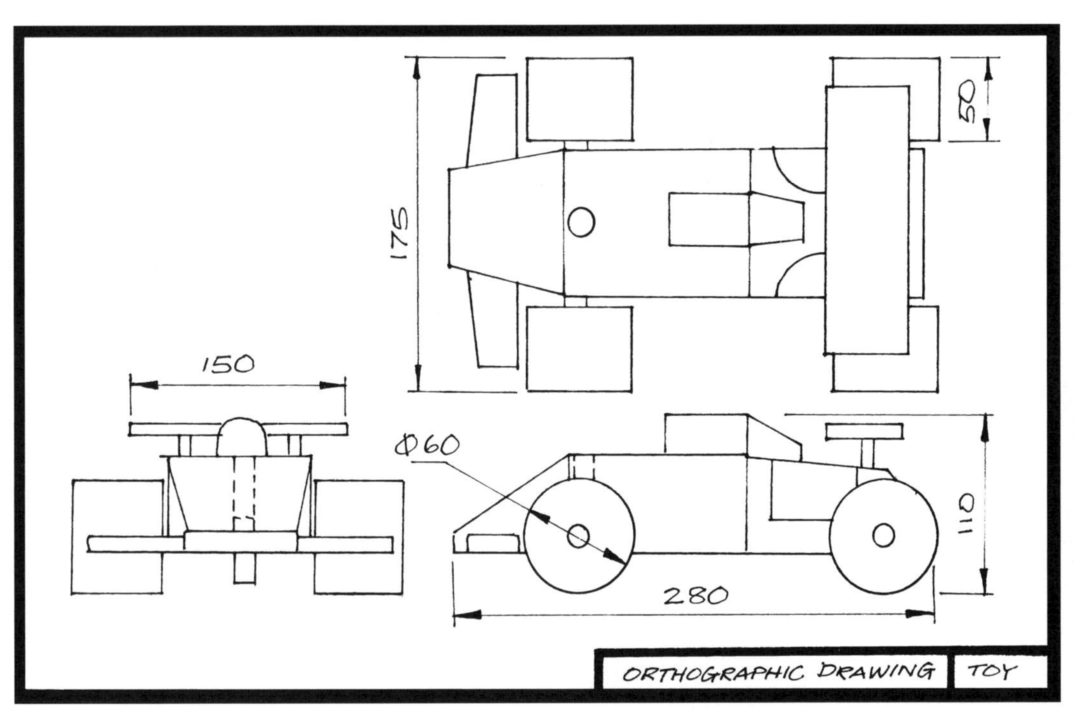

ICT 

Use a CAD program to produce your orthographic drawing.

IN YOUR PROJECT

Don't forget to make the mechanism the exact size of the mechanism in your toy. By doing this, you will also be able to work out the dimensions of the rest of the vehicle. Don't forget to use the anthropometric information that you gathered on page 81.

Construction (1)

Having decided on your design you need to think about how you are going to make it.

Work out which components will need to be joined together using fixings and fastenings, adhesives and/or joints.

Fixings and Fastenings

There is a wide variety of fastenings and fixings available, just look along the shelves in your local DIY shop! However, there are still rules to be followed when choosing what to use.

Nails

Nails are metal pins with heads. They provide a very quick method of joining wood and manufactured boards together. They are often used in conjunction with adhesives to add strength to the joint.

Nails are usually made of mild steel. Some, such as tacks, are tinned or 'blued' to prevent them rusting. They have small serrations on their shank, which catch on the fibres of the wood when they are in place. Once hammered into the wood, the heads are usually punched below the surface of the wood and the hole filled.

Name	Size	Uses
Round wire	20–150 mm	Cheap construction work: large head
Oval wire	12–150 mm	Fine work: heads can be punched below surface
Panel pin	12–150 mm	Thin sheet material, delicate work: head can be punched
Tack	6–30 mm	Short with large head, mainly for upholstery work
Staples	Various	Square for upholstery, round for wire or springs

Activity: joining two pieces of wood using nails

1 The correct type of nail should be chosen for the job. It should also be of an appropriate length to hold the two pieces of wood together.

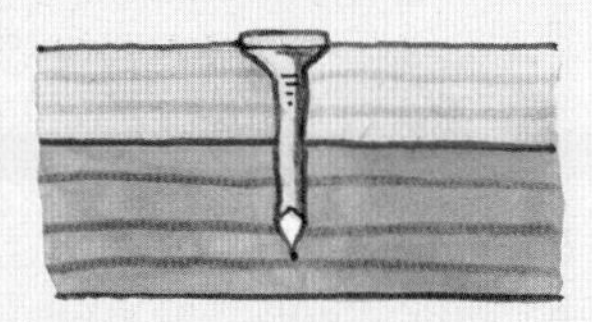

2 Start by knocking the nails into the top piece of wood in a straight line, so that you can just feel the tip coming through.

3 You may want to glue the surface of the other piece of wood at this point. Now place the two pieces of wood together and hammer the nails down through the top piece of wood and into the bottom one.

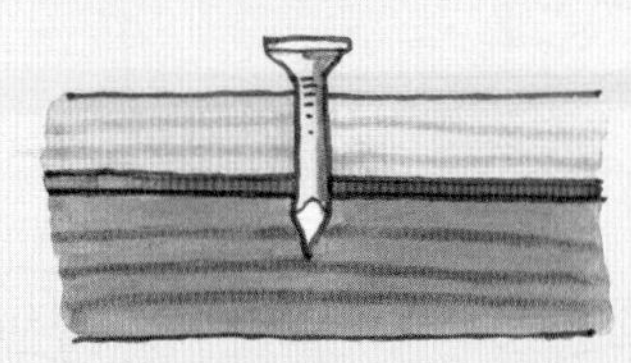

4 Now punch the heads below the surface and fill the holes.

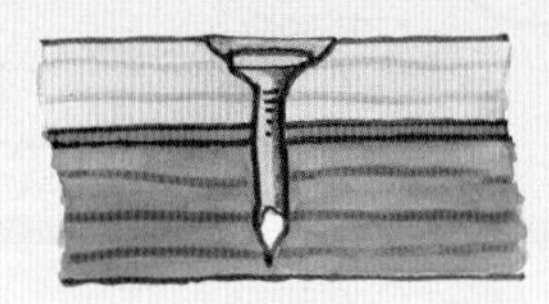

5 When removing a bent nail, always use pincers, rolling them onto a piece of scrap wood. Never try to knock the replacement nail into the same hole!

6 Dovetail nailing refers to hammering the nails into the wood at an angle to each other, making the two pieces of wood more difficult to pull apart.

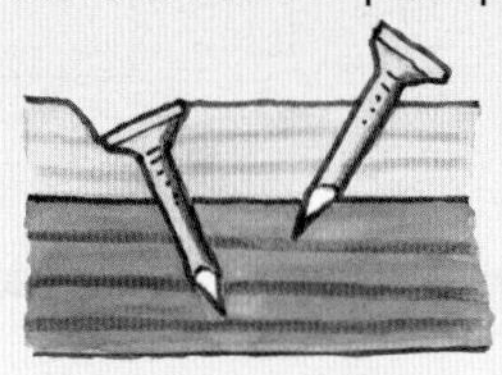

Screws

Screw size (or gauge) is determined by shank diameter. Gauge numbers (4, 6, 8, 10 etc.) stay the same whatever the length of the screw. They have better holding power than nails, yet they can be removed easily.

There are three different heads available:

- slotted head, which is the commonest
- Phillips head, which has a crossed recess and is less likely to slip than the slotted head
- Posidrive, which has a crossed recess head with a grip system.

Screwdrivers should always be chosen to fit the slot of the screw to avoid slipping and damaging the wood.

Type	Size	Uses
Counter-sunk head	6.5–150 mm	General purpose, needs a countersunk hole for a flush surface finish
Round head	6.5–87 mm	Fixing metal and plastic fittings to wood. Black Japanned finish for outdoor use
Raised head	8–50 mm	Decorative plating often applied, used to fix door handles, cupboard catches.
Twinfast	6.5–87 mm	For manufactured boards, it has two threads

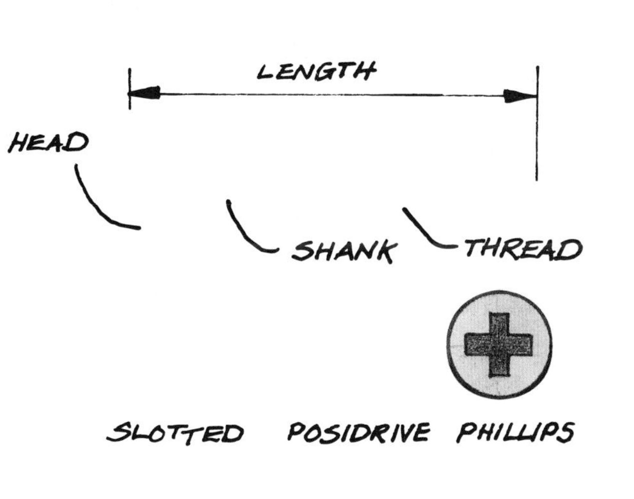

Activity: joining two pieces of wood using screws

1 First drill a hole in the top piece of wood or manufactured board. This is called a clearance hole and should be the same size as the shank of the screw. This hole should be countersunk if you are using a countersunk screw.

2 Now place the two pieces of wood together and drill a pilot hole into the lower piece of wood. It should be half the diameter of the clearance hole.

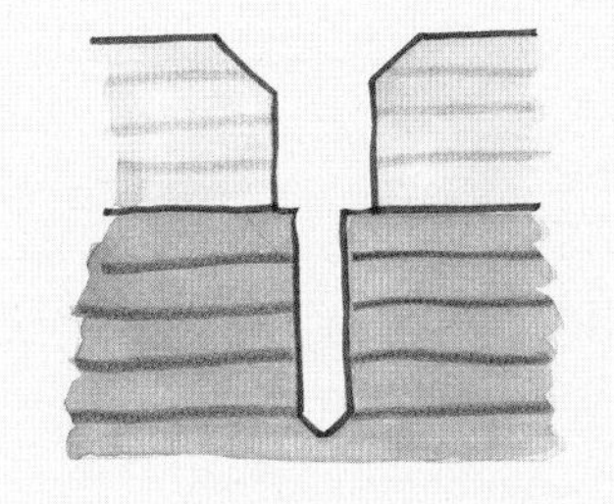

3 Select a screw of an appropriate length. Carefully screw the two pieces of wood together, keeping the screw at 90° to the top piece of wood.

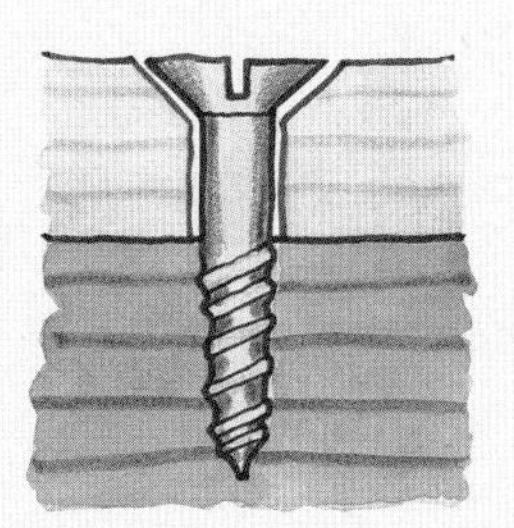

4 Sometimes screws do not grip in end grain. To rectify this, a plastic rawl plug may be inserted in place of the pilot hole. You need to ensure that they match the size of screw you are using. As you screw into the rawl plug it will expand against the sides of the hole and grip firmly.

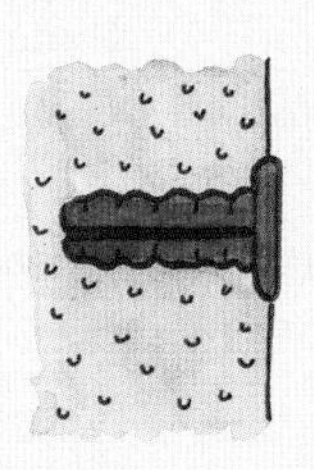

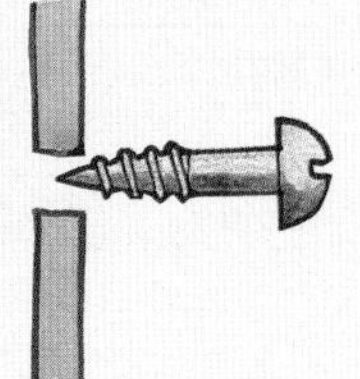

Construction [2]

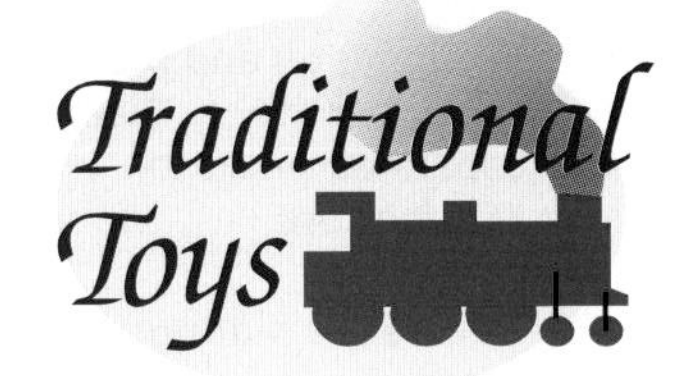

Adhesives

There is a wide variety of different **adhesives**. Each one has been developed for a particular situation. In the table below you will find details of the adhesive best suited to the materials you need to join.

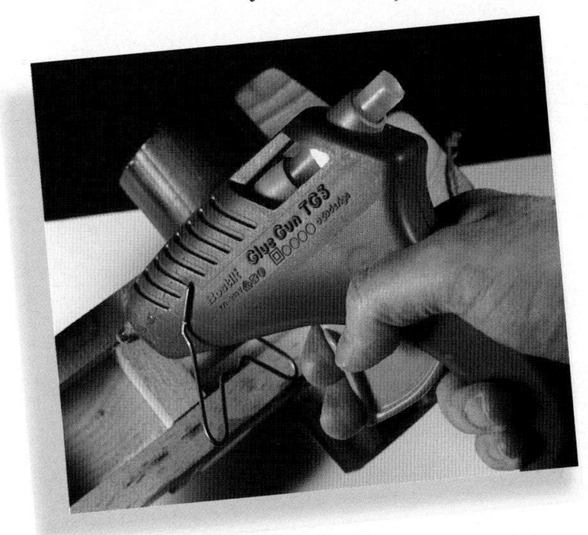

A glue gun can be used on most materials, but only on small areas.

Types of adhesives

Adhesive	Example	Appearance/Properties
PVA (Polyvinyl acetate)	Resin W	Thick, white liquid
Synthetic resin	Cascamite	White powder added to water, sometimes within separate liquid hardener
Epoxy resin	Araldite Rapid	Comprises resin and hardener which have to be mixed together in equal amounts
Acrylic cement	Tensol	Thick, clear liquid with unpleasant fumes
Contact adhesive	Thixofix	Thick, rubbery glue
Fabric glue	Copydex	White, rubbery glue
Rubber solution	Bostick	Clear rubbery glue
Hot glue	Bostick	Hotmelt

Which adhesive should you use?

Material	Wood	Metal	Acrylic	Melamine	Fabric	Rubber	Expanded Polystyrene
Wood	PVA						
Metal	Epoxy resin	Epoxy resin					
Acrylic	Epoxy resin	Epoxy resin	Epoxy resin Tensol				
Melamine	Contact	Contact	Contact	Contact			
Fabric	Contact PVA	Contact	Contact	Contact	Copydex PVA		
Rubber	Contact	Contact	Contact	Contact	Contact	Contact Rubber solution	
Expanded polystyrene	PVA	PVA	PVA	PVA	PVA	PVA	PVA

IN YOUR PROJECT

- How many parts of your design will you have to join together?
- What methods will you use?
- Will adhesive be strong enough or will you need to use nails or screws?

KEY POINTS

- Use the correct adhesive for the materials to be joined
- Make sure that you follow the manufacturer's instructions for use.

Joints

There are a variety of **joints** you can use to hold your design together. You will need to choose the most appropriate method. Think carefully about:

- ▷ how strong it needs to be
- ▷ how difficult it is to make accurately
- ▷ how much time you have got.

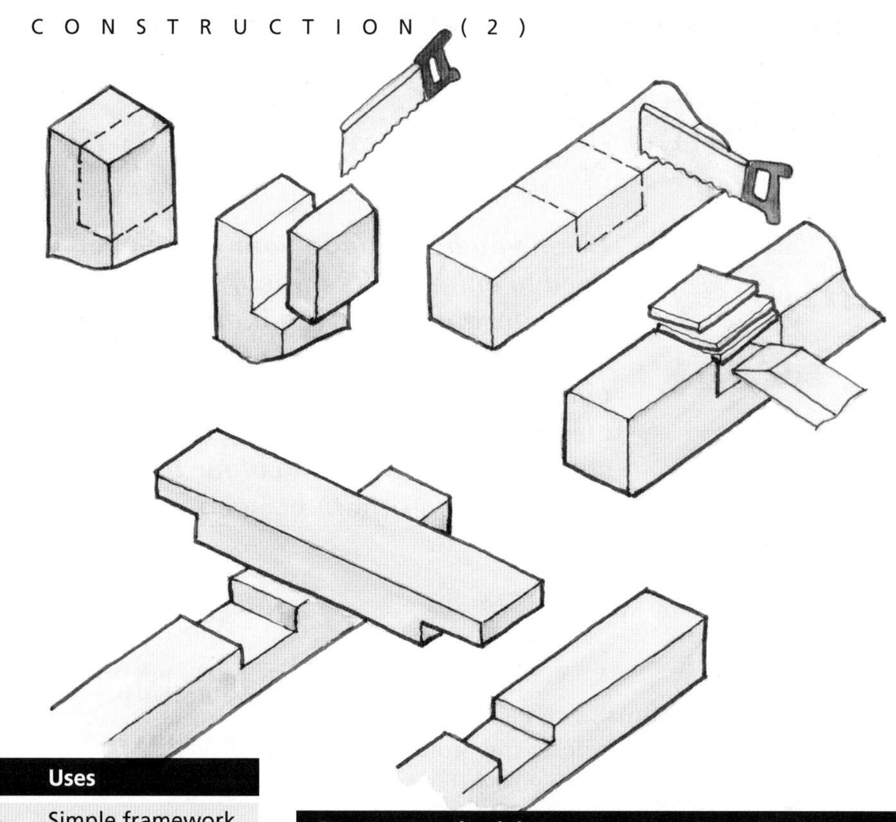

Corner joint	Uses
Corner-halving	Simple framework
Dowel joint	Simple joint with some strength
Bridle joint	Strong framework (e.g. doors) Difficult to cut
Cross halving	Frameworks without corners
Dovetail	High quality joint for cupboards and drawers

Box construction joint	Uses
Butt joint	Simple joint, glued together
Dowelled butt joint	Needs careful spacing
Mitre joint	Picture-frames – needs great accuracy
Housing joints	Shelves and partitions
Comb (finger joint)	Drawers. Very strong
Half-lap joint	Simple drawer joint

Fine Finishes

When you have made your design you will need to enhance its appearance and make it more durable by giving it an appropriate finish. You may decide to use paint or to bring out the natural beauty of the wood by using a clear or coloured varnish. There are also wood stains available which colour the wood while allowing the grain to be seen.

Preparation

Before you can apply any finish you need to prepare the surface of the wood. It must be clean, free from nail and screw heads (except where you have used these as decoration) and smooth.

All traces of pencil markings and glue must be removed. It is important that you remove excess glue with a damp cloth before it dries and that you use glass paper to remove pencil markings, not an eraser. Both glue and the rubber from an eraser can prevent the finish which you apply to your work from adhering to the surface.

Staining or Colouring

A wide variety of different **stains** is available. Some simply add colour to the wood while others stain and add a polyurethane finish at the same time. The technology of wood finish is rapidly changing and there are stains available which completely cover the surface of the wood, like a paint, yet still allow it to breathe.

Remember that the colour on the tin will not necessarily give the same result on your wood or manufactured board, so try some out on a scrap of material before you apply it to your toy.

Directions for use are given on the product.

Abrasive papers

There are two main types of **abrasive papers**:

- glass paper (particles of glass stuck onto paper)
- garnet paper (made with particles of garnet).

If used properly they can give an excellent smooth finish onto which your protective finish may be applied.

If used inappropriately however they can cause scratches and remove or spoil any fine shaping which you have given to the wood.

Abrasive papers come in different grades (see table) and should be used in the order coarse to fine as the imperfections are gradually removed. The abrasive paper should be wrapped around a block of cork or wood and rubbed in the direction of the grain. This prevents scratching. If the grain lifts when paint, stain or varnish is applied, you will need to rub it down again and reapply the finish.

Grade	Glass paper	Garnet paper
Coarse	3	2
	2½	1½
	S2	1
	M2	½
Medium	F2	0
	1½	2/0
Fine	1	3/0
	0	4/0
	00	5/0
Extra fine	Flour	6/0

Fillers

Sometimes it is necessary to fill wood, for example where you have used screws or nails during construction. Some lead-based fillers are still available and for safety reasons it is important that these are not used on toys.

Fillers come in the form of a paste and are applied with a palette knife or rag. They are then left to dry before rubbing down with abrasive paper. Sometimes fillers are referred to as 'stopping'.

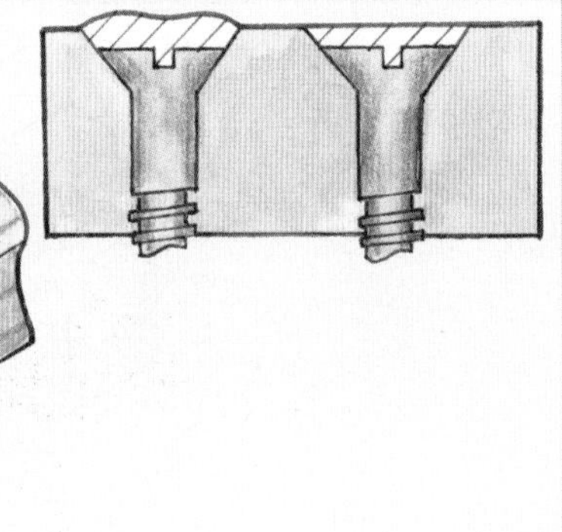

Paint

There are several different types of **paint** finish available, from high gloss to completely matt. Paint provides a colourful, durable finish, especially for softwood and there is a wide range of colours available.

Oil based paints

This paint is based on oil and coloured pigments. Silicone is often added to help surface cleaning and polyurethane to increase surface hardness. It can be thinned with white spirit. Gel or non-drip gloss paint is also available. Wood is usually sealed, using a primer and then an undercoat before the final coat of colour is applied. Again, improvements in paint technology have meant that one coat products are available which make this sequence unnecessary. They are usually suitable for both indoor and outdoor use.

Polyurethane paint

This paint is tough and scratch resistant. It can withstand high temperatures and is ideal for toys and furniture. It is usual to apply two or more coats, sanding down any raised grain between each application, but again, developments in paint technology have resulted in one coat products being available and you should read the manufacturer's instructions. It may be applied with a spray or a brush. If a spray is used, a mask must be worn. It is an ideal finish for MDF (medium density fibreboard).

Enamel paint

This paint is the traditional model makers paint. It is ideal for adding fine detail to your work. It is oil based and brushes must be cleaned with white spirit. It should not be confused with glass enamelling which is a finish for metal and is a totally different process.

Varnishes

Shellac

Many schools have shellac varnish available for you to use. This is made from crushed beetles and is one of the oldest ways of finishing wood. It is used in the the traditional and very skilled method of finishing furniture called French polishing.

If you use shellac you will need to apply several very thin coats, rubbing the surface down between each application. Although it is not the ideal finish for a toy, it does give an attractive appearance if applied carefully.

Polyurethane varnish

This is essentially liquid plastic. It is a synthetic resin which dries to a durable finish, having the same properties as polyurethane paint. It can be applied with a brush or spray. If using a spray, a mask must be worn. It should be applied in several thin coats, and left to dry thoroughly between each application.

In order to obtain an excellent finish, rub wax into the surface along the grain, using fine wire wool. Wait until it is completely dry. This results in a very smooth, silky finish.

Masking

It is possible to paint or varnish areas with great accuracy using a mask. This is a piece of paper, or a length of masking tape, which protects the areas which you do not want to paint or varnish with a particular finish. You can then protect the areas you have painted or varnished and apply a different finish to the rest of the toy. Be very careful when you remove the mask not to remove some of the paint finish with it.

■ ACTIVITY

Try out a number of different finishes on scraps of wood and manufactured boards. You will be able to judge more accurately which you should choose for your toy.

IN YOUR PROJECT

- Choose the finish which has the properties which make it suitable for your material and design.
- Prepare surfaces carefully before applying finishes.
- Remember that some finishes are toxic.

KEY POINTS

- All surfaces must be properly prepared before painting or varnishing.
- Try out a finish on scrap matetial before applying it to your product.
- Finishing may take as long as making.
- Follow the manufacturer's instructions carefully.

Planning the Making

Check that you have all of the information that you need to start making. Then you can start to plan your sequence of work. This needs to be organised so that all the time available to you is usefully employed. Don't wait around waiting for glue to dry when you could be applying a finish to another part of your toy.

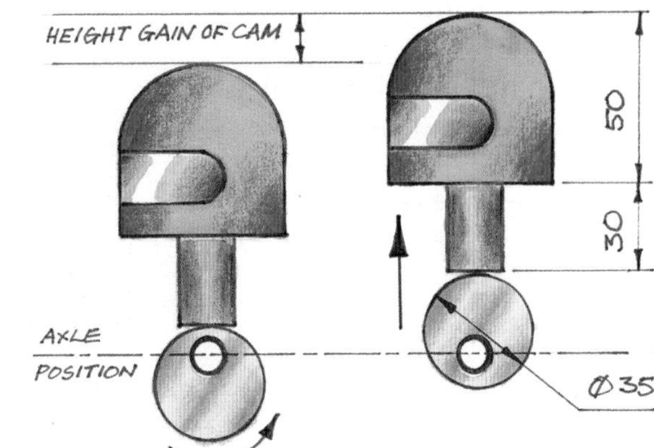

Ready, Steady...

Make sure that you have an accurate drawing of the toy and its mechanism, together with the model of the mechanism. Your drawing should clearly show how the mechanism will work, including the 'up' and 'down' positions of any rotary cams which you may be using.

Prepare a list of the materials which you will want to use and make sure that they are available, including the components which you will need for the mechanism. Check on availability of adhesives, nails and screws and finishes.

You need to make a flow chart of all the processes which you will need to go through and allocate time for each one. Be as realistic as you can be. Make sure that the time you allow adds up to the time available!

IN YOUR PROJECT

- Organise your work so that there is some overlap between the different stages. Can you be making the mechanism while the glue is drying on something else?
- Make sure that you use accurate measurements.
- How can you ensure a high standard of finish?
- Highlight hazards and list safety precautions that need to be taken, particularly if working with spray paint.

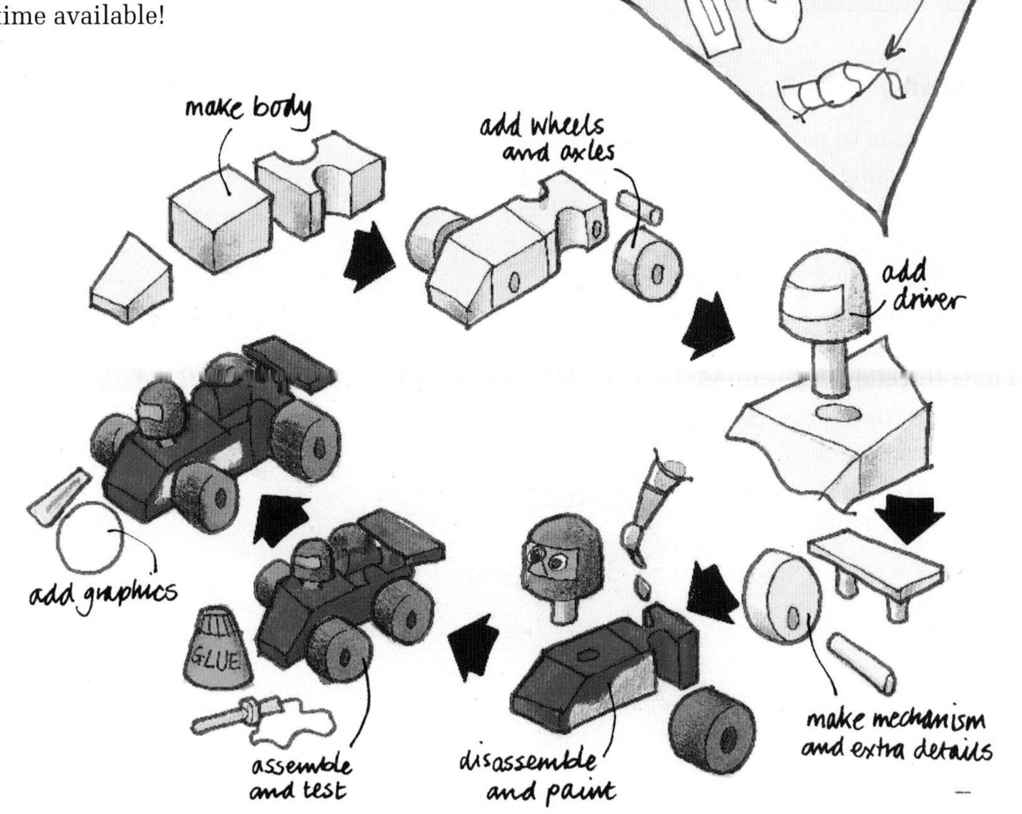

Testing and Evaluation

Look back at the specification which you wrote down in response to the letter which the Once A Tree Managing Director sent to you. How well does your solution meet the requirements of the specification?

- Does it have moving parts?
- Does it have a simple mechanism?
- Does it conform to the British Safety Standard for toys?
- Does it use renewable resources?

Remember to write the 'Green' Report

You will need to devise some tests to confirm that it conforms to the British Safety Standard for toys (see page 92).

How else could you test the success of your design?

'How well does the mechanism work?'

User survey

The best way of evaluating the play value of the toy is to ask some children to play with it and tell you what they think about it.

Perhaps you could arrange to take your toy to a playgroup. Ensure that it conforms to the safety standards before you take it. Keep a careful eye on the children to make sure they are using it properly, and that it is not being roughly handled!

Planning for manufacture

How would you set about making 1,000 toys? How would you organise the production? How might templates and jigs help? How might the design need to be changed to make batch production quicker, easier and more cost-effective?

Final Evaluation

Write your **evaluation** report. Remember to mention process and product:

- both the good and the bad points about the processes you used
- how you overcame problems and changed your plans
- how you would change the processes if you were to make your toy again
- how closely the design fits the specification
- how well the mechanism works
- what the results of the user survey were
- how you could improve the design of the toy.

Examination Questions

You should spend about one and a half hours answering the following questions. To complete the paper you will need some A4 and plain A3 paper, basic drawing equipment, and colouring materials. You are reminded of the need for good English and clear presentation in your answers.

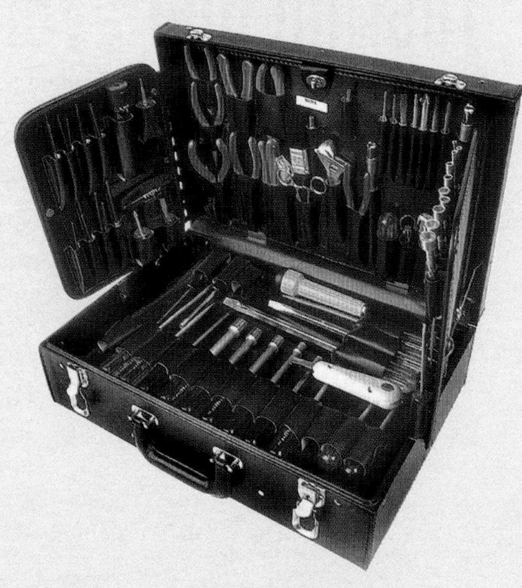

1. This question is about research into existing solutions. See pages 76-77. *(Total 4 marks)*

a) Name two ways of carrying out a survey of children's toys. *(2 marks)*

b) You have surveyed a group of Nursery children playing with plastic and wooden toys. You have found that plastic toys are more popular than wooden toys. How could you best present the information? *(2 marks)*

2. This question is about designing. See pages 80-81. *(Total 4 marks).*

a) What is anthropometrics? *(2 marks)*

b) How could it be useful when you are designing a wooden toy? *(2 marks)*

3. This question is about mechanisms. See pages 86-87. *(Total 8 marks)*

Look at the illustration below of a charity collecting box that is shaped like a dog. When a coin is placed on the dog's tongue and a handle is turned, the coin is tipped into the box.

a) Re-draw the mechanism, adding the parts that will tip the tongue and slide a coin into the dog collecting box, when a handle is turned. *(4 marks)*

b) Next show where the handle will go on the mechanism. Make sure that it is in the right place to turn the mechanism. *(2 marks)*

c) What is this mechanism called? *(2 marks)*

4. This question is about mechanisms. See pages 84-85 and 90-91. *(Total 6 marks)*

a) What is the name of the type of motion used by the mechanism in Question 3? *(2 marks)*

b) What is the name of the type of motion used by the tongue? *(2 marks)*

c) Which class of lever is the dog's tongue? *(2 marks)*

5. This question is about safety. See pages 92-93. *(Total 6 marks)*

a) Explain what safety checks you would carry out on a child's wooden toy. *(4 marks)*

b) There is at least one safety problem with the design of the collecting box. Explain one of them. *(2 marks)*

6. This question is about materials. See pages 82-83. *(Total 8 marks)*

There are two main categories of wood, softwood and hardwood.

a) Name two types of softwood and give one use for each. *(4 marks)*

b) Name two types of hardwood and give one use for each. *(4 marks)*

7. This question is about materials. See pages 82-83. *(Total 8 marks)*

a) Give two reasons for choosing European redwood for making a child's toy. *(4 marks)*

b) Give two reasons for choosing beech for making furniture. *(4 marks)*

8. This question is about construction. See pages 100-101. *(Total 12 marks)*

You have decided to use a countersunk screw to join together two parts of a product.

Using notes and sketches, describe the three stages of joining two pieces of wood together using a countersunk screw. *(4 marks per stage)*

9. This question is about construction. See pages 102-103. *(Total 4 marks)*

Look at the illustration above.

a) Which construction joint would be best to use to make a wooden photograph frame with an acrylic design? *(2 marks)*

b) Which adhesive should you use? *(2 marks)*

10. This question is about finishes. See pages 104-105. *(Total 6 marks)*

Explain how you would prepare the surface of a wooden frame and add a finish. *(6 marks)*

11. This question is about environmental concerns. See pages 78-79. *(Total 4 marks)*

a) What does biodegradable mean? *(2 marks)*

b) Give an example of a biodegradable material that you have used. *(2 marks)*

12. This question is about planning for making. See pages 106-107. *(Total 10 marks)*

Look at the drawing of the simple wooden toy shown below. Draw a flow chart of the five stages of making of this toy. *(10 marks)*

Total marks =80

Project Four: Introduction

Combining Colours and Textures (page 116)

Designing for Manufacture (page 140)

Quality Counts (page 138)

Furniture companies manufacture products to suit an increasing demand for low-cost furniture that can be quickly and easily assembled at home. This type of furniture is supplied in a flat-pack form.

Many* flat-pack *products are objects which will be used for storage in the home or office. What ideas can you come up with for new designs of* self-assembly *storage items?

MARKET RESEARCH REPORT

Manufacturers produce a range of models to satisfy different markets. Companies are keen to spot a gap in the market, i.e. a product model or variation which is not well supplied by other manufacturers.

The market for quick self-assembly products is vast. There are many opportunities to launch new products. We have identified a particular need for simple storage devices which look modern and exciting. These would be aimed at people in their teens and twenties living in one-room, a small flat or their first home and looking for unusual shapes, colours and textures to express their independence.

A manufacturing company has asked you to develop designs for a new product that will save space in the home or office, and be sold to the customer in the form of a flat-pack. The product must be designed for manufacture in large quantities to meet expected demand.

You are also asked to produce a leaflet to show clearly how your design can be assembled easily by the customer.

Wherever possible use ICT to help you design and make your item of flat-pack furniture (see pages 10–11).

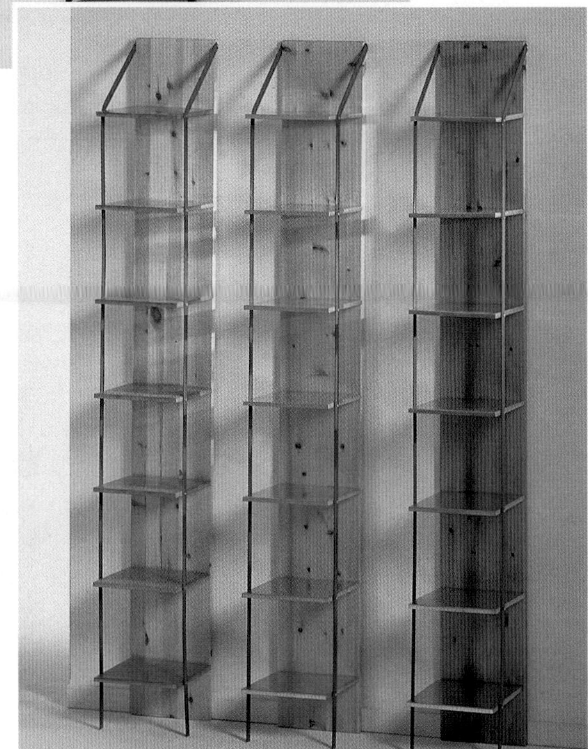

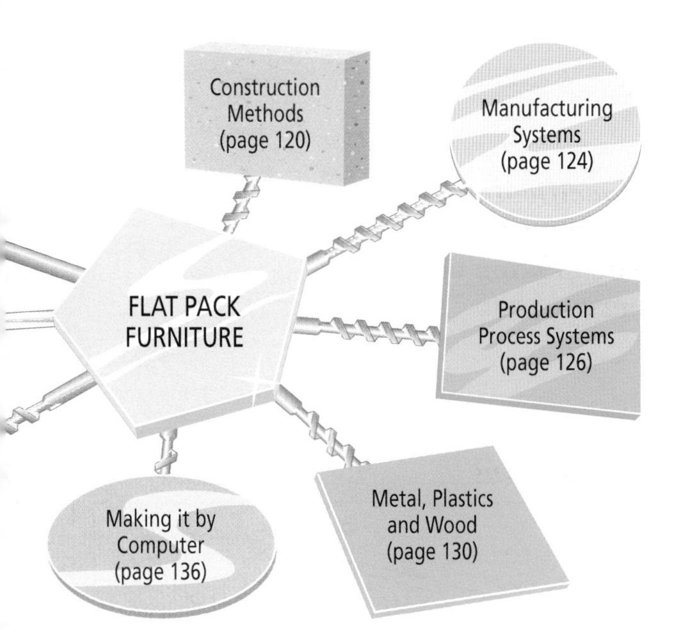

Clarifying the Brief

You are asked to come up with ideas for a simple, low-cost storage device which:

- can be assembled in the home or office
- will be made from MDF, metal and/or plastic
- will be of a modern, eye-catching appearance
- will be simple to mass-produce.

First Thoughts

What ideas have you got for possible storage items?

Brainstorm a list of items at home that need to be stored.

Sketch some initial ideas. Remember that the shapes you draw should be simple, unusual and colourful.

ACTIVITY

- Have you had self-assembly furniture bought for you from a local furniture or DIY store?
- What was it ?
- How did you get it home?
- Did you have any problems in assembling the product?

To find out more about contemporary furniture for the home go to:
www.ikea.com
www.habitat.net

Planning the Investigation (1)

Before you can prepare a design specification you will need to use a number of methods of research. These will need to include looking at existing solutions and making an ergonomic study.

Start collecting information about things that need to be stored and existing ways of storing them. You will need to do more than just cutting pictures from the leaflets and magazines that you have collected. It is important to record all the information you have gathered and where you obtained it from. Present your findings on a number of A3 sheets.

In-store

As well as having self-assembly furniture on display, some furnishing stores and DIY centres offer a free catalogue of products. Try asking for further information on materials, fixings, costs, etc.

At home

Most homes will have an example of the type of product you are investigating. Look around the lounge, the bedroom, the kitchen, etc., and you will probably see many items which need storing.

Devise and undertake a survey of the people in your group who have an example of a product that was bought from a local store in the form of a flat pack.

Existing Solutions

Designers often develop new ideas by studying existing products, and where, when and how people use them. They look for things which work well, and for features they might be able to improve on. Sometimes this involves taking products apart, but more usually they use their eyes to take a product apart visually.

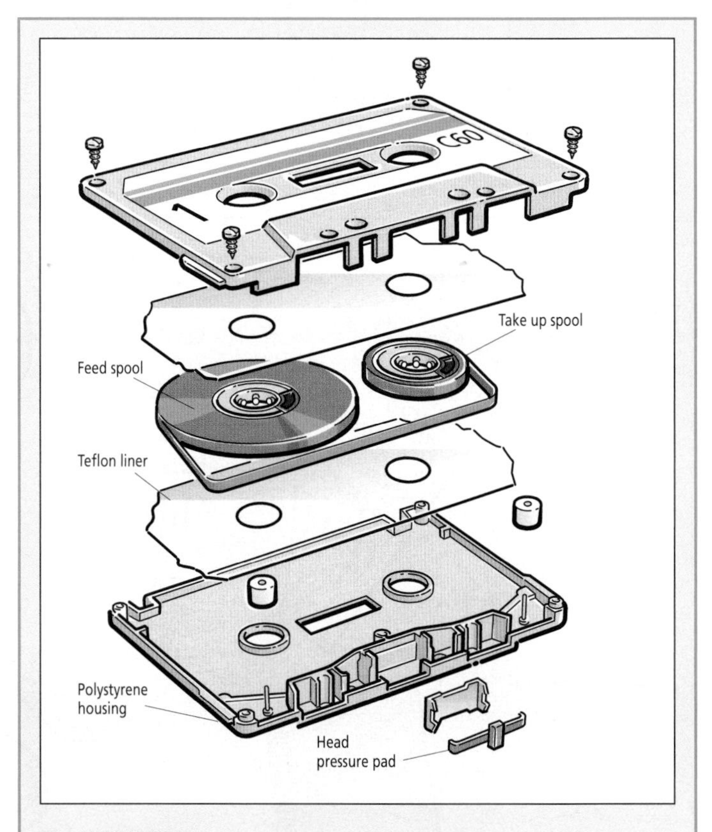

■ ACTIVITY

Carefully disassemble an audio cassette tape which is no longer needed.

- What does each component do and how does it do it?
- What materials are used for each piece?
- How does each part fit with the others?
- What properties do the materials need to have to make the cassette safe to use?
- What statements about the use, performance, materials, appearance and manufacture of the product do you think might have been in the original design specification?

Make separate detailed drawings of the various parts. You may like to use tracing paper or a light box to plan the most effective position for each part in the **exploded drawing**.

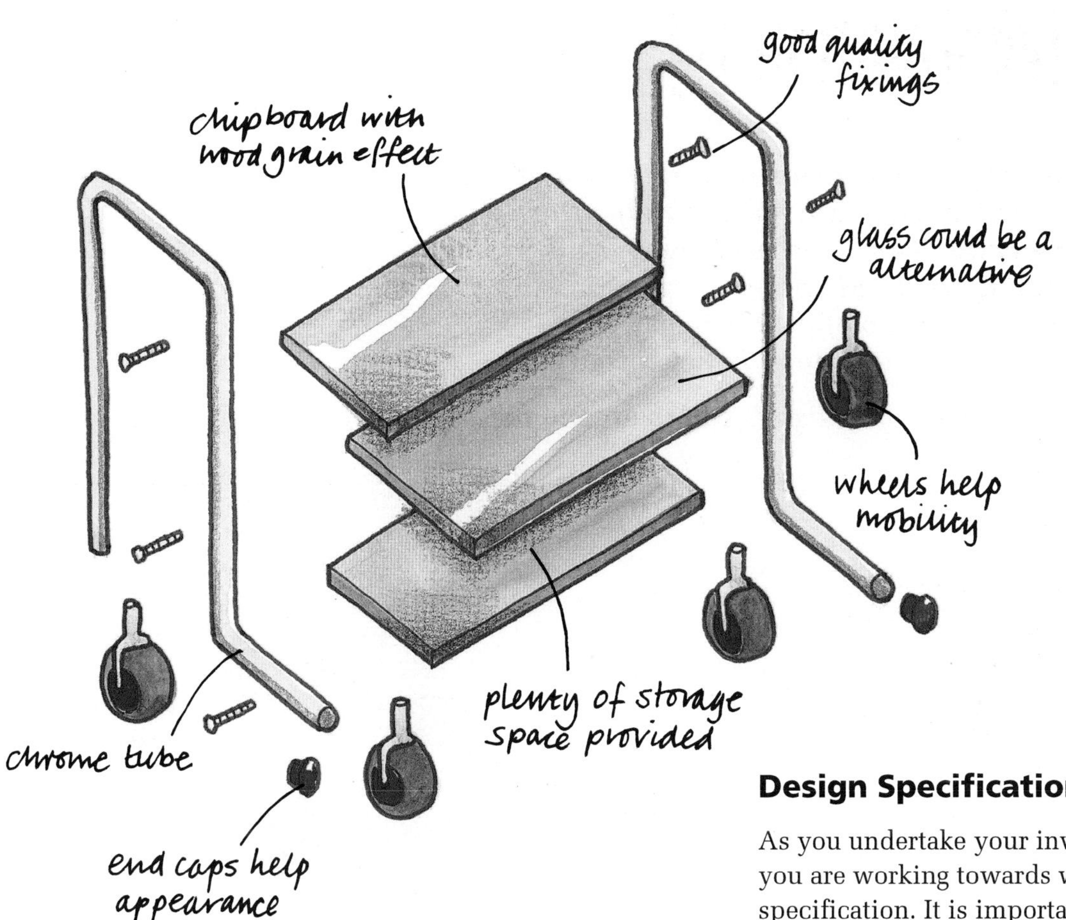

■ **ACTIVITY**

Identify an item of self-assembly furniture and examine it in detail to see what it is made from and how it has been fixed together.

Design Specification

As you undertake your investigation, keep in mind that you are working towards writing your design specification. It is important that you will be able to cover the key features associated with the manufacture and distribution of flat pack furniture. These will need to include:

- ease of assembly by the customer
- the total number of units likely to be made
- the size of batches, if appropriate
- the requirements for assembly instructions.

Specification for flat-pack CD storage holder

- The unit must hold at least 10 CDs.
- There must not be more than six components, excluding fittings.
- The unit must be suitable for manufacturing in batches of 10,000.
- It must not take more than five minutes to assemble with the aid of a standard screwdriver, and no other tools.

Making It in Quantity

At this early stage you may be thinking more about designing and making a one-off product. Later, however, you will need to consider the issues involved in planning a production line for the making of your design. Different manufacturing processes and materials can be used according to the numbers to be produced.

Some designs prove to be difficult to make quickly and easily in quantity. This is often because of their:

- shape
- arrangement of components
- material and finish.

Towards the end of the project you will probably need to change your final design to make it more suitable for manufacturing. Think about how this might be done. What might you need to change?

Investigation [2]

Product designers need to go to a lot of trouble to make sure their products are a pleasure to own and use. This includes designing them to be:

- ▷ ***easy and comfortable to use***
- ▷ ***easy to understand how to use.***

***This aspect of design is called* ergonomics.**

Things which are small and fiddly to open, or difficult to work out how to operate have not benefited from an ergonomic study. It is important to design for the situation in which a product will be used and for the capabilities of potential users, particularly if they have special needs (e.g. children, the physically and visually handicapped, and the elderly).

Sometimes relevant information is already available in books. If not, it may be necessary to set up a series of tests and experiments to obtain the data which is needed.

Sensory evaluation

Prepare a coloured drawing of the product. Add notes to describe the following ways in which people experience it:

- ▷ what it looks like (its colours, shapes and patterns)
- ▷ what it sounds like (any noises it makes, either as part of its use, or unintentionally)
- ▷ what it feels like (the texture of the surfaces of the materials it is made from)
- ▷ what it smells like (the materials it is made from and/or when in use)
- ▷ what it tastes like (if appropriate).

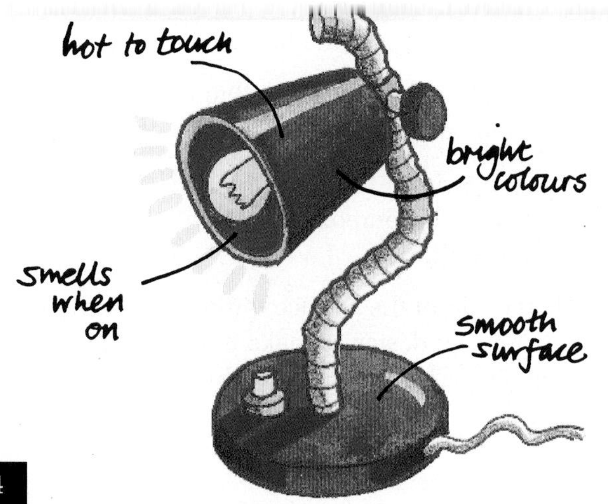

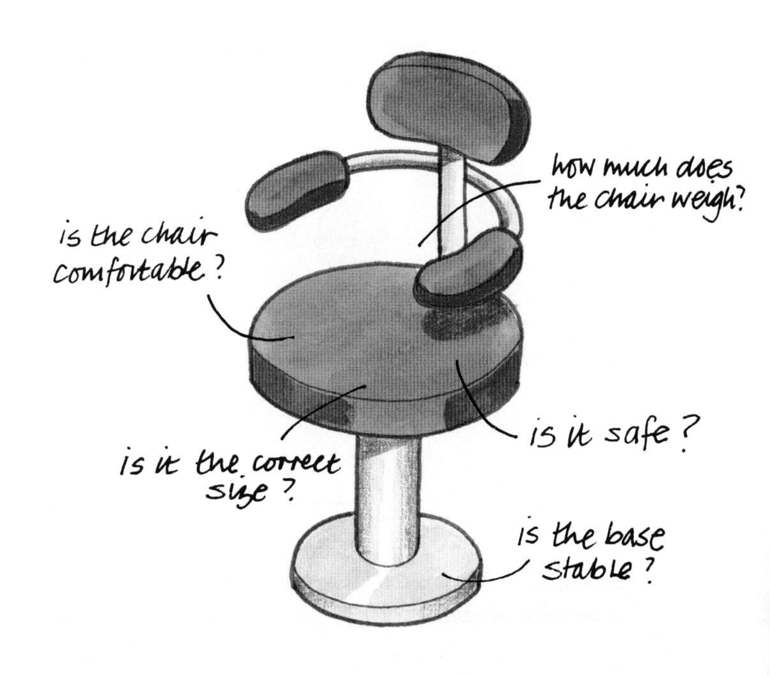

■ ACTIVITY

Choose either:

- ▶ an item of furniture in your home or school, such as a chair, table, cupboard, etc., or
- ▶ a small electronic product you use often, such as a personal stereo, camera, alarm clock, bedside light, etc.

How well designed is it? What might be improved? Think carefully about things like comfort, safety, strength, weight and size:

- ▶ what task is being performed (e.g. storing a CD)?
- ▶ how easy is it physically to use the product (e.g. lifting, moving)?
- ▶ are any instructions, information and control devices placed in the best arrangement and sequence?

Personal Products and Spaces

Make a detailed ergonomic study of someone you know. It could be a friend or relation who lives close by, or another member of your family – but it can't be yourself! You will need to find out more about:

▷ the physical characteristics of the person you choose
▷ their likes and dislikes
▷ a space they work or relax in
▷ the items they need to store
▷ the products that enable them to store things
▷ items which are not presently well stored.

IN YOUR PROJECT

As a result of making a study of someone's storage requirements you might identify a suitable product to design for your project.

KEY POINTS

Designers use ergonomics:

- to help develop the overall shape of a product
- to decide the arrangement of the main component parts, and controls and displays
- to define requirements for people with special needs.

Physical capabilities and limitations

Take some general measurements of the person you are studying. As well as overall sizes, take some more unusual measurements which describe their capabilities and limitations:

▷ how far they can stretch up when standing and when sitting
▷ how far they can reach up when standing and when sitting
▷ their manual dexterity – how well they can perform delicate or fiddly tasks
▷ visual abilities – how good their eyesight is at different distances in different lighting conditions.

Likes and dislikes

▷ Ask your subject about the colours, shapes, patterns, textures they like best, and those they dislike.
▷ Get them to choose something they have recently purchased and to describe why they like it.
▷ Can they name an item they really enjoy using, and one which always makes them feel frustrated or dissatisfied?

Work/play space

▷ Take some measurements of the space.
▷ Are things which are used together placed together?
▷ Is the lighting adequate?
▷ Is the heating and ventilation satisfactory?
▷ Are there any safety hazards?

Storage

▷ Make a list of all the things stored in the space.
▷ Note all the various devices used for storage (e.g. cupboards, drawers, racks, trays, etc.)
▷ Which items are not stored very successfully?

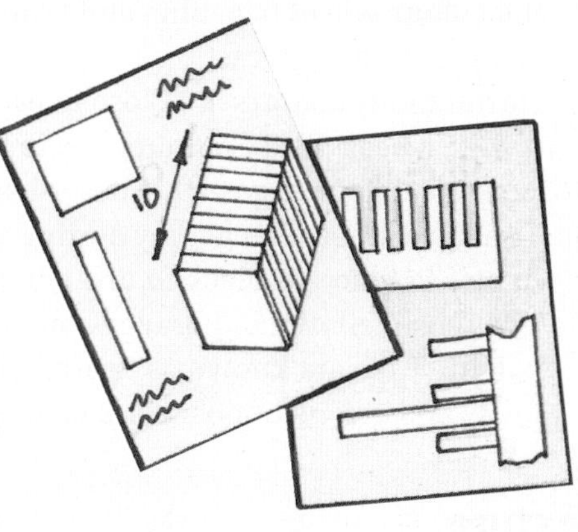

Combining Colours and Textures

Combinations of colours and textures of materials can be a source of harmony and contrast. These elements help make a product look interesting and distinctive and create dramatic effects.

Colours can be in harmony or contrast with each other. They can alter the way we feel by creating different moods.

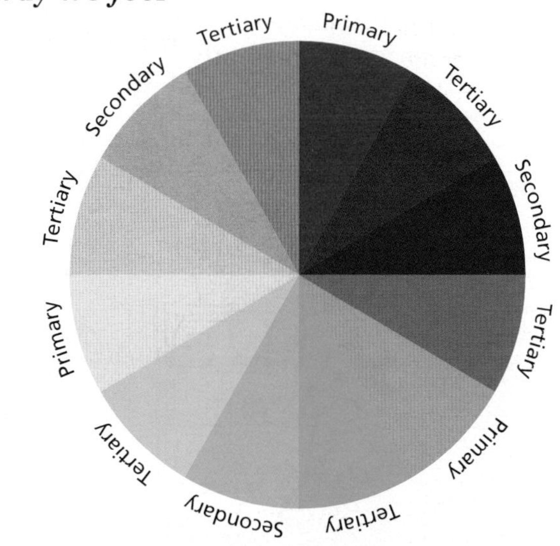

The Colour Wheel

- The **primary colours** are red, blue and yellow.
- The **secondary colours** are purple, green and orange. These are produced by mixing two primary colours together.
- The **tertiary colours** are produced by mixing a primary colour with an adjacent secondary colour.
- **Complimentary colours** are opposite each other on the **colour wheel** (e.g. blue and orange). These colours create contrast
- **Harmonious colours** are close to each other on the colour wheel. These colours create harmony.
- A single colour (e.g. red) is described as the **hue**.
- The hue can be changed by adding white to make it lighter (a **tone**), or black to make it darker (a **shade**).
- The reds, yellows and oranges on one side of the colour circle are known as 'warm' colours, while the blues and greens opposite are called 'cold' colours.

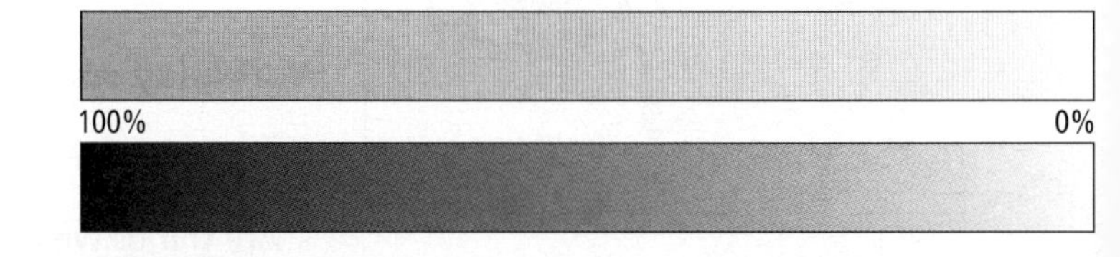

Texture

Products have texture as well as colour. The texture of a surface can change the way a colour looks. Patterns and textures can be very regular and geometric or very free-flowing and natural looking.

It is always important to experiment with samples of different colours, patterns and textures using the materials which will finally be used.

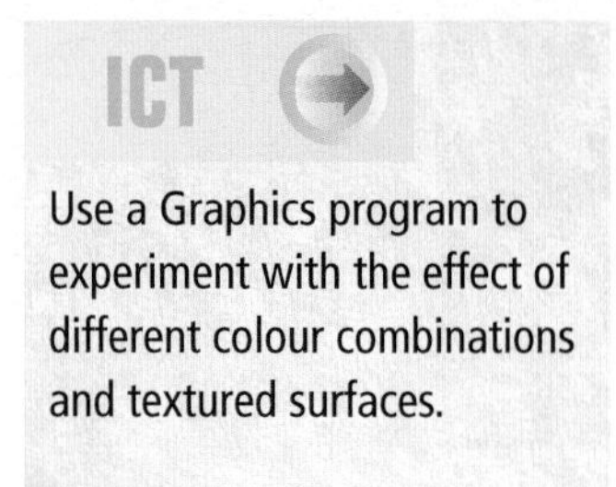

Use a Graphics program to experiment with the effect of different colour combinations and textured surfaces.

ACTIVITY

In a group, assemble six different products, all of which have a variety of colours and textures. They could include packaging as well as items made from woods, metals, plastics, textiles, ceramics and/or glass.

For each object select three words which you think describe it well. Compare your results with others in your group.

COOL	WARM
PASSIVE	ACTIVE
FEMININE	MASCULINE
NATURAL	ARTIFICIAL
EXCLUSIVE	POPULAR
INDIVIDUAL	CORPORATE
TRADITIONAL	MODERN
CLASSICAL	ROMANTIC
RETROSPECTIVE	FUTURISTIC

Memphis case study

Memphis is the name of a group of Italian architects and designers who created a series of highly influential products during the early 1980s. Their work included designs for domestic products, furniture, lighting, fabrics and environments. They set out to challenge the idea that these objects had to follow conventional shapes, colours, patterns and textures.

The colours they use contrast the dark blacks and browns of most European furniture.

Improbable combinations of colours, patterns and shapes draw cultures together – Burger King red, next to the violet of a West African textile, and a bright sunlight spray-can yellow.

The influence of the Memphis designers can be seen everywhere. Look out for examples of the similar use of their colours and patterns, particularly in restaurant areas, in graphic products and in fabrics and furnishings.

Images from Studio Aldo Ballo and Studio Azzurro

KEY POINTS

- Colours are either primary, secondary or tertiary.
- Colours can be used to create harmony and contrast.
- Colour is always affected by light and texture.

IN YOUR PROJECT

► Consider the potential market for your product. Which ranges of colours and textures and patterns would be most appropriate?

Modelling Ideas

Designers use models to help them visualise the shape, form and proportion of a product.

It gives them the opportunity to handle and rotate the model to determine whether it will work in reality. These models can be 3D, or generated by computer.

To find out more about 2D and 3D CAD programs go to:
www.bentley.com
www.adobe.com

Some 3D models are used to help designers develop their ideas. These are often called **prototypes**, **mock-ups** or **test-rigs**. Other models are used to communicate design ideas to a client, or to present them to the public. Making models for public presentation is covered on page 122.

You need to be very clear about the purpose of your model:

- who is it intended for – yourself or the client?
- what information do you hope to gain from making it?

When you have worked this out you will be able to choose the appropriate scale and the most suitable materials.

If you haven't already done so, you need to decide on the storage item you are going to design. The next stage is to make a simple prototype and then go on to think about selecting appropriate materials and methods of production before finalising your design.

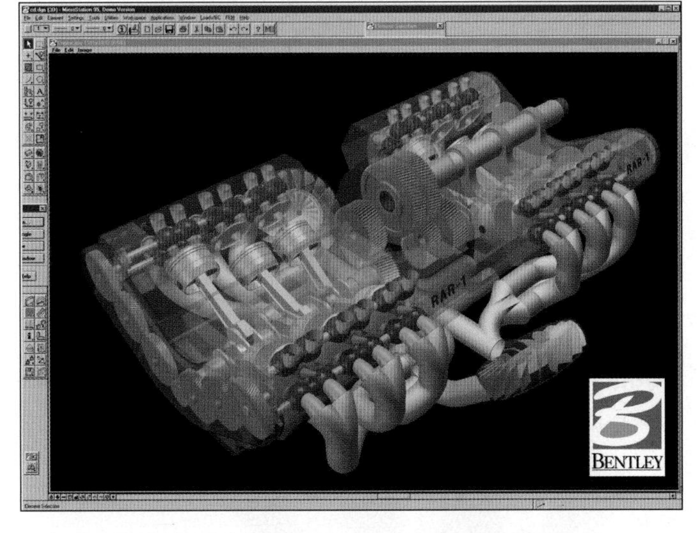

Computer Aided Design (CAD)

CAD systems make it easy to create a picture of a new design, and then to change the way it looks on the screen. This means that new ideas or variations on an idea can be tried out and evaluated much more quickly than if they had to be redrawn each time.

There are a number of different types of CAD programs available. Some are more concerned with the visual appearance of a product, others with technical detail.

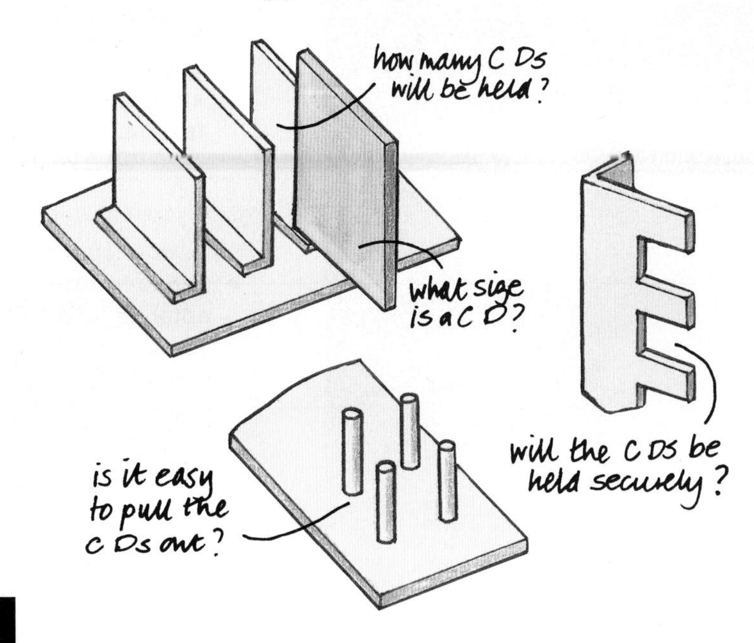

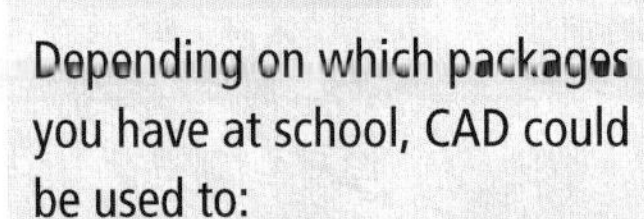

Depending on which packages you have at school, CAD could be used to:

- create a 3D surface-textured drawing of your idea
- produce orthographic and detail working drawings
- work out the arrangement and number of flat-pack units that can be obtained from the original un-cut manufactured board.

Methods of representation

There are three main ways of representing an object on a computer:

- **wire frame models** mean that the object is represented by a series of lines. This image can be enhanced by removing lines that are hidden.
- **surface modelling** can be added. The surfaces of the object are represented by colour, shading and texture to give a stronger sense of 3D form.
- **solid modelling** where the drawing is based on geometrical shapes which can be mathematically analysed.

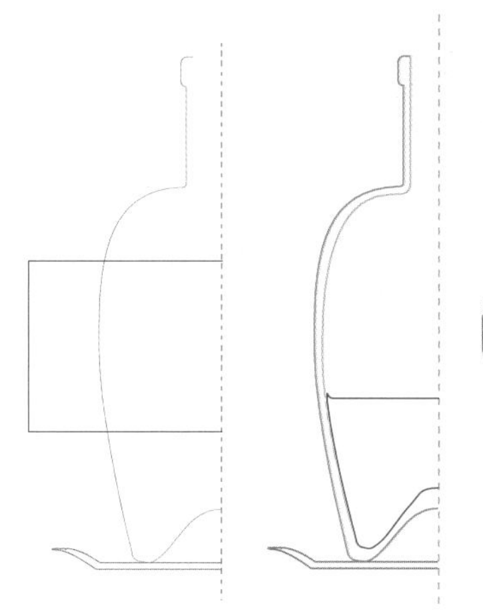

Transforming a CAD model

Most CAD programs have certain common facilities. Drawing tools produce lines which can be adjusted for width, shape and colour and can be set to produce curved lines as well. The designer is also able to access a library of existing images which can be added to a design.

It is very easy to **transform a CAD drawing** or image by copying, rotating and mirroring different parts of it. These functions are usually available on an on-screen menu.

Rapid prototyping

Rapid prototyping is a term used to describe the quick and easy production of a real three-dimensional appearance model of a design using data from a computer-model which has been created on screen.

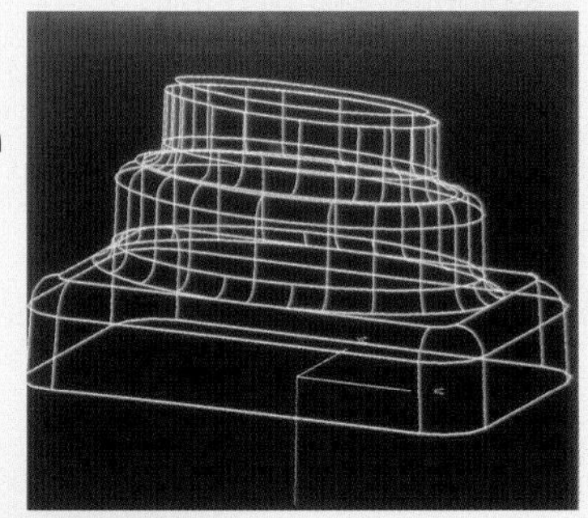

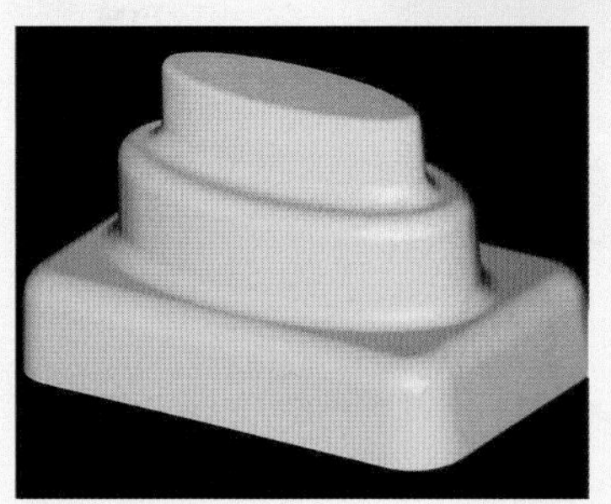

Instead of progressively removing materials from a solid block until the shape emerges, the rapid prototype is built up in a series of highly accurate cross-sectional layers made in epoxy resin. This avoids the need to make complex production tools. (See also page 137.)

KEY POINTS

The key advantages of using CAD are:

- it speeds up the process of design development, and as a result a greater variety of designs can be produced.
- an improvement in the quality of design, because the computer can more accurately simulate and produce accurate information about how the design will behave in different operating conditions.
- any changes can be made quickly and communicated throughout the team working on the project.
- design information which has been generated can be stored in a database or file and can be quickly and easily retrieved at a later date.

Construction Methods

How will your design fit together? What fixtures and fittings are available? What different sorts of handles and door hinges can be used? How can shelves and drawers be added?

Fittings and Fixings

Joining different materials together is a major consideration for the designer. Selecting the best method can be crucial to the success of the product. In some cases the method of assembling or joining can be seen as an added feature to the overall appearance of the product.

Press covers are designed to fit the heads of Phillips or Pozidrive screws and can be bought in a range of colours.

Mirror screws can also be used as these allow a chrome plated dome to be added to the screw head.

Selecting the best method of joining

Ready-made fittings ease the joining of components not only during manufacturing but also for the user.

Whilst developing your ideas, it would be helpful to try to test a range of fittings to see which are most appropriate.

When you are trialling these fittings it will be important that you work accurately. The positioning and alignment of each part will determine the success or failure of the assembly.

In certain circumstances a product can be assembled using a special coarse-threaded screw. These self tappers are especially effective in chipboard, although great care must be taken to ensure these screws are not positioned too near to an edge: this can swell or fracture the chipboard.

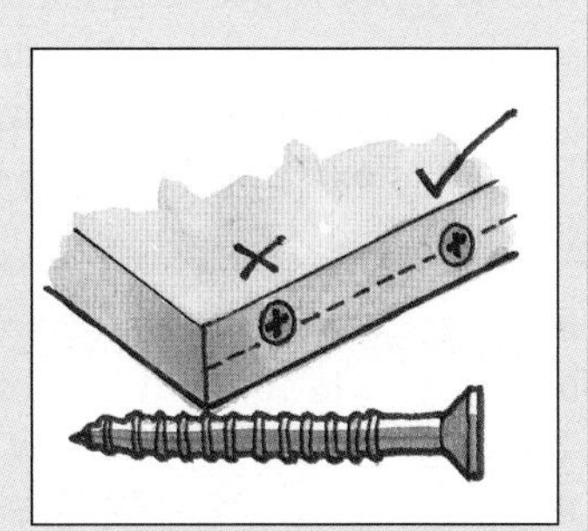

It is advisable to drill a pilot hole smaller than the core of the screw before completing the joining of two boards.

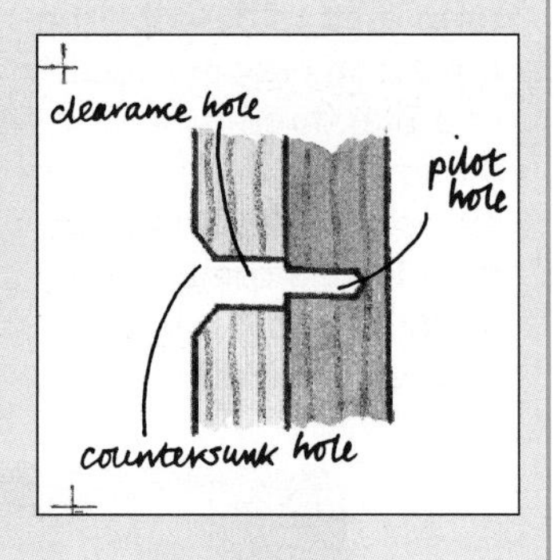

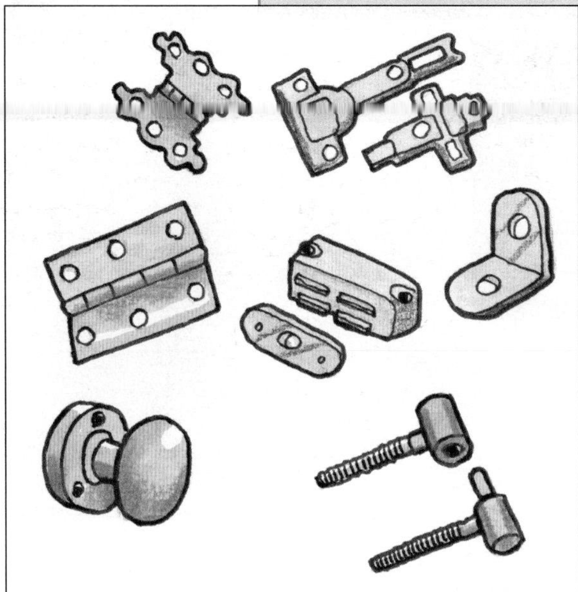

■ ACTIVITY

Some fittings will need to be purchased from specialist retailers whilst others can be bought from a local DIY store. Go and visit a DIY store and look at the wide range of fittings and fixtures available.

Fitting a shelf

Some fittings are used to allow individual preference in the design. For example the positioning of a glass or acrylic shelf in a storage system can vary to allow for specialist equipment to be housed neatly. These shelf supports can be a feature and are available in a range of materials which include brass, chrome-plated steel, nickel-plated steel, zinc and nylon.

Where a shelf is required and the fitting needs to be hidden a two part connecting system can be used. Here a plastic housing is pressed into an 18 mm blind hole in the shelf. This is then fitted to the zinc die cast supporting part screwed to the upright. The conical interior pulls the shelf tightly against the upright. The supporting part locks automatically in the housing by means of integrated tabs.

Some fittings use an Allen key. This is a piece of hexagonal metal rod, bent at 90° which fits into a hexagonal hole in the top of a bolt. This is seen as adding a feature to the design and no attempt is made to hide this type of head.

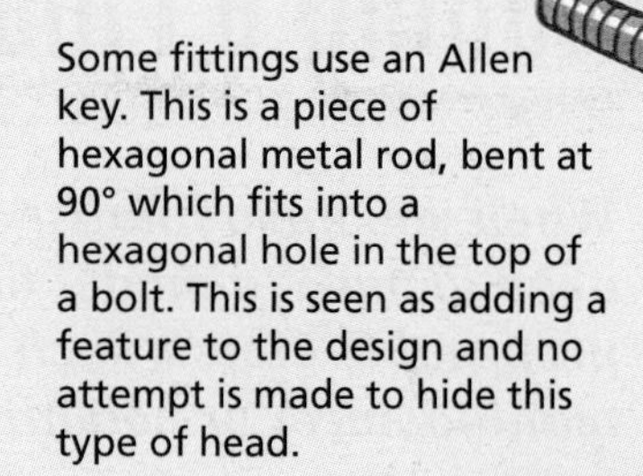

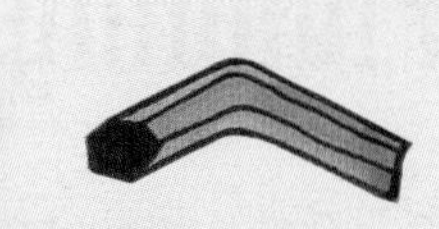

Some manufacturers machine joints that aid the assembling process and combine the use of traditional joint construction with readily available fittings.

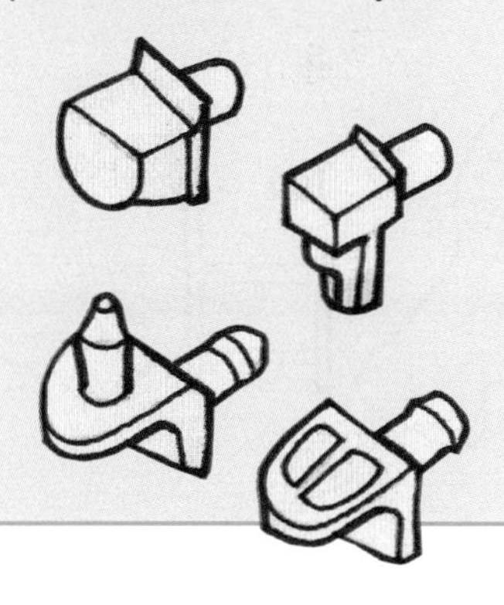

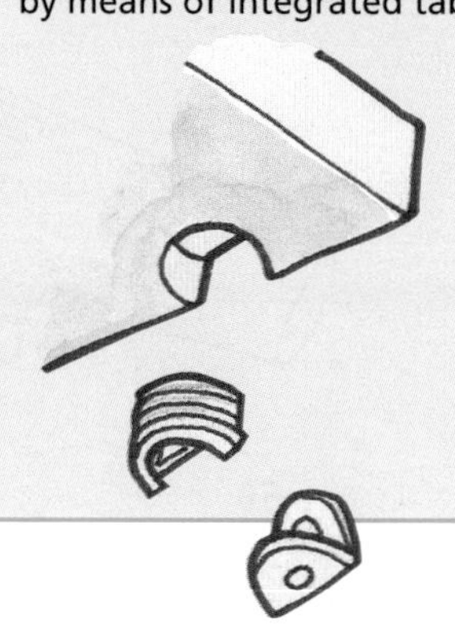

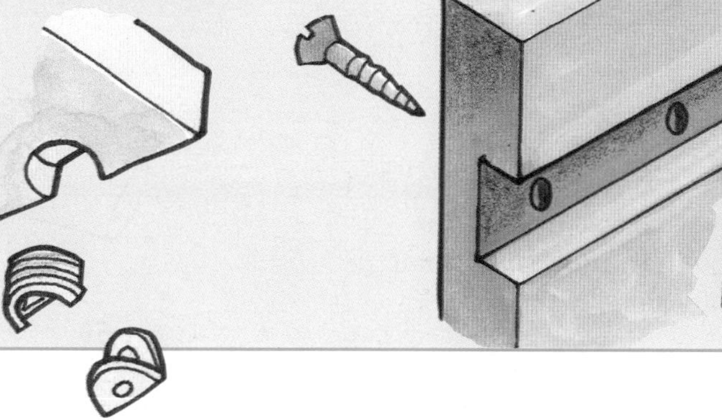

Drawer Construction

You may decide to include a drawer or set of drawers in your design. This used to be a demanding task that proved difficult to complete to a high standard. Using the right fittings, however, it can be accomplished reasonably easily and to a high degree of accuracy.

Some of the more simple connecting fittings consist of a threaded pin and a knock-in socket which incorporates a steel spring disc.

Firm connection by a push-fit makes it particularly suitable for flat pack furniture. This method of fitting provides economic advantages as it does away with the time consuming work required for traditional or dowel joints.

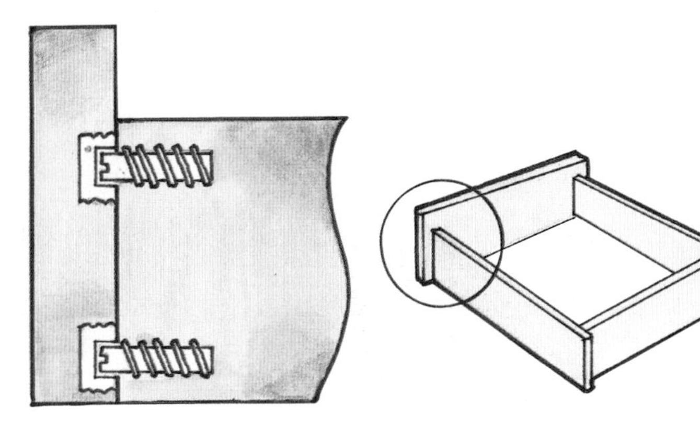

Runners

There are several ways of making a drawer run smoothly. Most require a high degree of accuracy, not only in making the drawer, but also in ensuring that the carcass is square and vertical and that the runner is at right angles to the upright.

In your design you may consider adding a simple runner to the upright, allowing the underside of the drawer to run on it

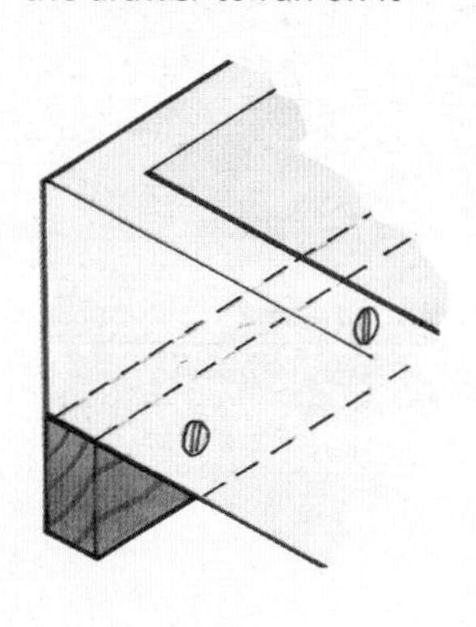

Alternatively you could make a groove in the side of the drawer with the runner positioned accurately on the upright.

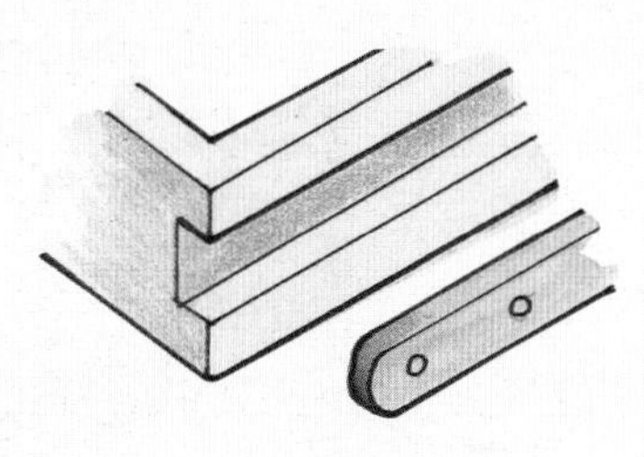

A third option is to add a two piece ready-made unit which is fitted to the drawer side and the upright. Although this is a more expensive solution it does ensure the drawer runs well and has a built in mechanism to ensure the drawer does not pull out completely.

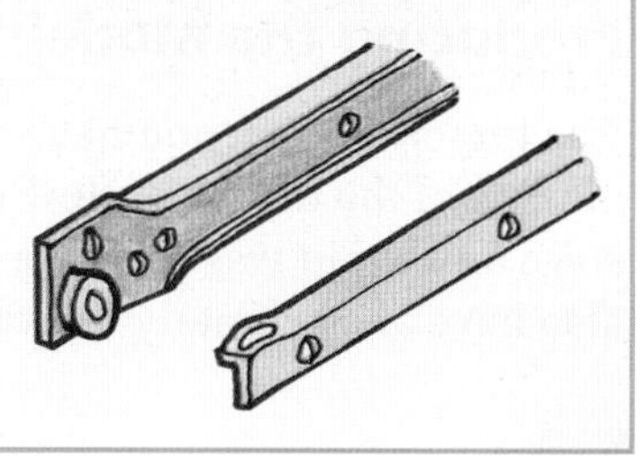

Planning and Making a Final Product Model

When you have finalised your design idea you need to make a model to communicate your proposal to others, such as a client, manufacturer or potential customer.

Model making is a specialised skill in its own right. It is important that you try to make a product model as accurately as you can.

Scale

One of the first decisions is what size to make the model. Think carefully about how much fine detail you want to include. Here are some of the more commonly used **scales**:

1 A small component – 2:1 (i.e. twice as big)
2 A desk-top or hand-held item – 1:1 (full-size)
3 A piece of furniture – 1:10 (1/10th size)
4 Part of a room – 1:50
5 A house and garden – 1: 100

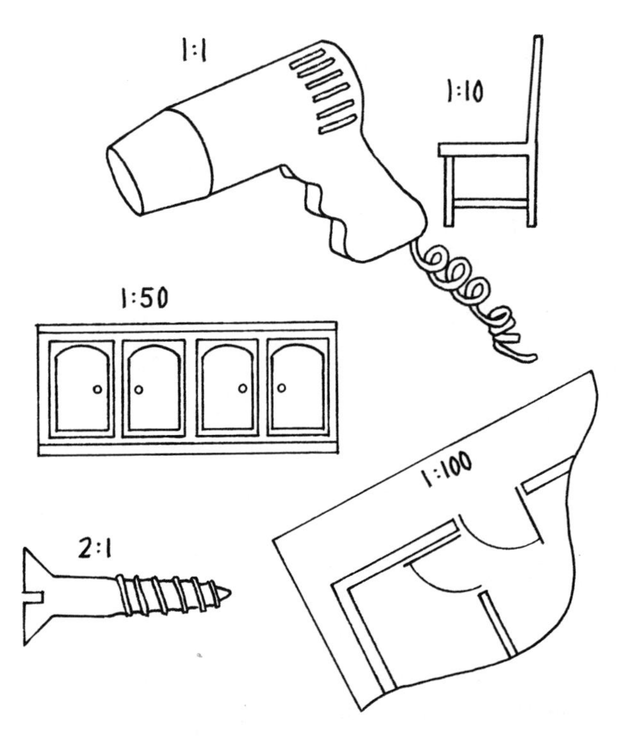

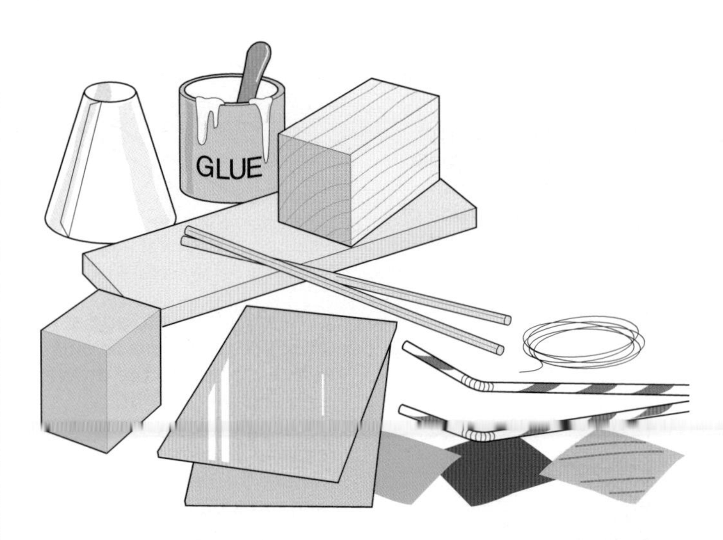

Materials

Most 3D models use different materials to those which would be used for the final product. You may find that for making furniture you can use flat materials such as foam board, card, thin MDF or odd bits of wood or discarded objects. The important thing is to use those materials that can be shaped, cut and finished to a high standard.

Producing the Model

It is important that you have a clear route through the making of the model. A flow chart can help plan each stage and avoid wasting time and materials. You should also have an accurately drawn **workshop drawing**.

Finishing Touches

Great care should be taken to ensure the final product model is well finished.

MDF

Most self-assembly products are made using either chipboard or medium density fibreboard (MDF) and possibly hardboard and plywood. Because of its open and coarse end grain, the edges of chipboard need to be 'finished' in some way. The most popular is to apply an edging strip of either a hardwood or a plastic laminate.

However it is possible to select a material such as veneered-faced chipboard, which does not require a finish. Black ash, mahogany, pine and coloured plastic laminates are all readily available.

Remove unwanted glue and sand the material to size at each stage. You may wish to radius edges: this will require skill.

Painting

Some models may need painting.

▷ A number of light coats are better than one heavy coat.
▷ Lightly rub down the surfaces with wet and dry paper before the final coat is applied.

It is important to use a primer and to avoid using different types of paints on one surface: Cellulose and Acrylic can react with one another and act like paint stripper. Some paints react badly with materials such as polystyrene, so do some tests on scrap materials first.

If you are able to use spray paints then the process can be a lot easier. However, you will need to ensure the room is well ventilated and that all safety requirements are observed.

On the edge

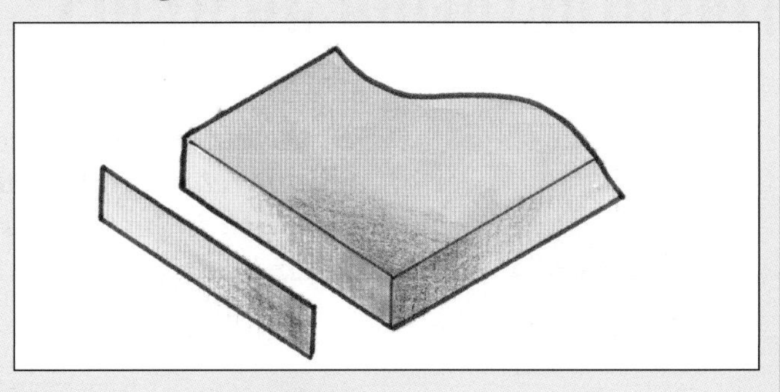

Some boards are edged in length and only need facing on the short sides or where a joint is exposed. This can be difficult and accurate marking out is required if the thickness of the veneer is to be levered when the joint is complete. These edges tend to be square and do not add much to the design of the product.

The increasing use of MDF does allow for slightly more creativity in the design as the edges can be shaped or formed using a planner or router.

The equivalent industrial process is to complete these operations on a router table where the machinist will cut a number of the same units to create a batch.

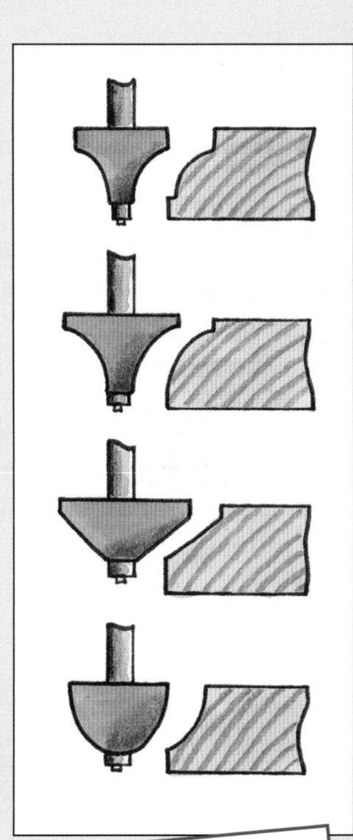

SAFETY FIRST!

Using a router
You may be allowed to use an electric router to enable you to cut joints and shape edges to a high degree of accuracy. As with any electrical tool, safety regulations must be followed. Only use a router under the close supervision of your teacher.

Testing

You should use the final product model for testing and analysis. Let someone else give their opinion of the design. They should comment on the shape and form, the sizes of each component, and the detail of extras such as drawers and shelves.

What you learn will help you improve your design as you go on to consider how it will be manufactured in quantity.

Manufacturing Systems

To design your product to be suitable for manufacture you need to know about the way in which industrial production systems are organised.

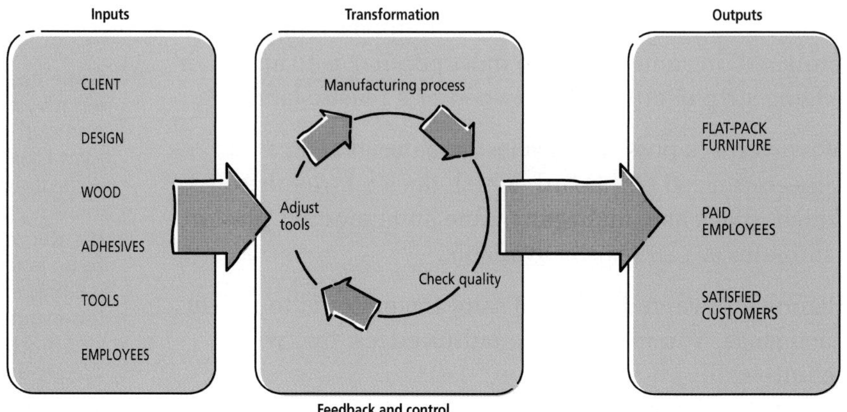

Systems and Control

A **production system** is an inter-connected series of events, materials and components. It is important for a designer to understand something about how such systems work together to make a product.

The role of a production manager is to establish how well a system is working and to find ways in which it could be improved, so that a product can be made more efficiently.

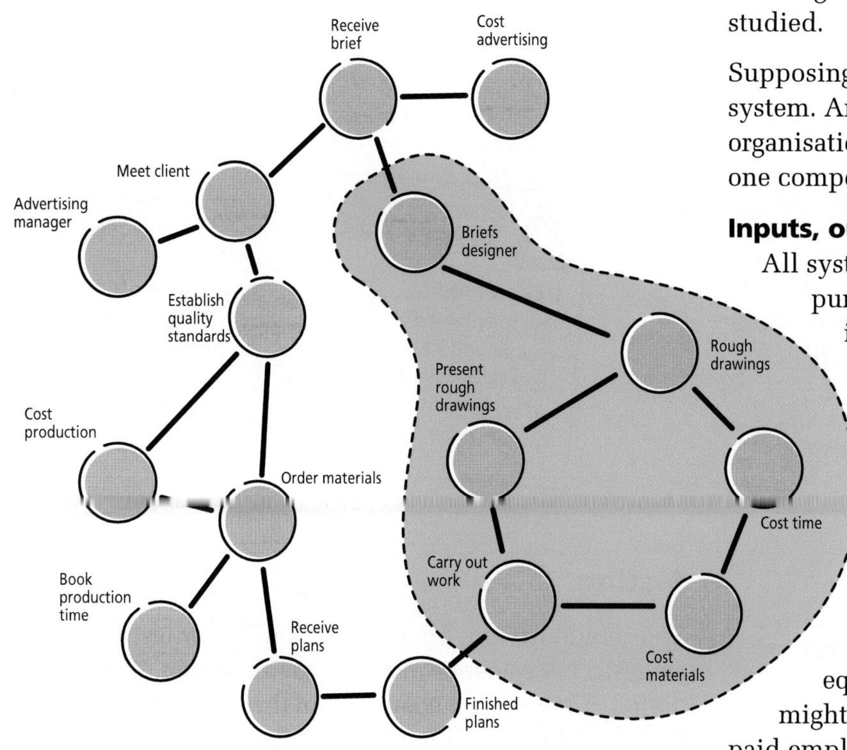

System boundaries

All systems are parts of an infinite number of larger and smaller systems. It is therefore essential to begin by defining the particular **boundaries** of the system to be studied.

Supposing you wanted to study a factory production system. Are you going to consider the whole organisation which it is a part of, or just how efficiently one component is being made?

Inputs, outputs, and transformation processes

All systems have **inputs** and **outputs**. The main purpose of a system is to change or transform the inputs into outputs.

Most systems have many different sorts of inputs and outputs. The first stage in analysing a system is to identify the inputs, outputs and the **transformation processes** involved.

Again if we use a factory as an example of a system, the inputs might include raw materials, employees, manufacturing, equipment, energy, etc. The transformed outputs might include items of furniture, waste materials, paid employees, etc. The factory is a system for bringing such transformations about.

It is also possible to analyse a system in terms of a sequence of events, e.g. collect raw materials, prepare and mark them out, cut them, etc.

Feedback and Control

Some transformation processes serve to maintain the **equilibrium**, or balance, of the system. Others work to improve the quantity and/or quality of the outputs.

It is possible to identify and analyse whether the various processes going on are maintaining the balance or attempting to improve quantity or quality.

When undertaking a systems analysis it might be discovered that the quantity or quality of the outputs are unsatisfactory in some way – too much waste material, or products which fail to meet the necessary quality specification and have to be thrown away.

As a result it may be found necessary to change the inputs, or to alter the process of transformation. This is known as **feedback**. The means by which the inputs or processes are changed are called **controls**.

The success of a system is judged by considering how well it transforms its inputs into outputs, and how well it is prevented from failing to work satisfactorily as a result of its feedback and control mechanisms.

In the factory you might be examining the efficiency of a machine which bonds two surfaces together with an adhesive. It might be discovered that the acceptable limits of the amount of glue used are plus or minus 10%. Below this figure the surfaces will not stick adequately, and above the limit the manufacturing costs increase unacceptably.

The question then becomes how to ensure the machine works within these limits?

As well as quality of product or service provision, systems analysis often focuses on achieving acceptable production times.

Undertaking a systems analysis

Analysing a system involves looking at a complicated situation and being able to identify some degree of structure and connection between the things which are going on.

STAGE 1

Identify and clearly state:

- the main inputs and outputs of the system.
- which specific inputs and/or outputs you are interested in investigating (i.e. which you suspect to be unsatisfactory in some way).
- the parts of the feedback and control operations which need to be investigated (i.e. which you suspect to be unsatisfactory in some way).

STAGE 2

The next step is to collect as much information as possible about the way in which the parts of the system you are interested in work. This may involve collecting factual information (e.g. how long or how often something takes to happen) and opinions (people's observations and comments). In particular you need to try to find out:

- what happens when things don't run smoothly, or in an emergency
- how the system is maintained over different periods of time.

STAGE 3

Finally, evaluate the extent to which:

- the transformation process is over complicated and wasteful
- the limits of the feedback system (i.e. the points at which mechanical adjustments or decisions for action are made) are too high or low
- suitable provisions have been made for emergency operation and maintenance
- the system can be easily modified to take account of later needs
- various changes in the original inputs to the system would improve its performance.

Production Process Systems [1]

What method of production will you be using? Most production processes are based on one of the following methods:

- **one-off production**
- **batch production**
- **mass production**
- **continuous production.**

Templates, jigs and moulds are often needed to assist efficient production.

Production Processes

One-off production

It might take several days or even weeks for a craftsperson to produce a single item. This is costly in terms of labour and materials, although it does result in a very high quality product.

Batch production

Where the tasks and manufacturing equipment are shared, a team of people can produce a larger number of an identical product, in less time than if each person worked on their own. Working in this way they can also respond quickly to changes in market demand and switch to making a different design. This is known as batch production.

Mass production

If a number of workers are organised in the workplace on a production line then they will be able to make identical products very quickly, eight or more hours a day for weeks or months on end.

Although this significantly reduces time and costs, all production quickly halts if there is a problem, and changing the line to make a different product can take a long time.

Continuous production

Continuous production is when the production process is set to make one specific product twenty four hours a day, seven days a week, possibly over periods of many years. This occurs in some areas of food manufacturing (e.g. bread), the production of chemicals, steel, energy, etc., where it would take a long time to stop and re-start the production process.

Most production processes involve a mixture of these methods. Some parts might need to be individually or batch produced, while others will be run off continuously.

Different types of manufacturing equipment are needed for the different processes. Some require special purpose tools made to suit a particular product, while others require basic machines with parts which can be changed and reprogrammed when needed.

Standardised Design

In many industries a basic standardised design is used to maximise on the economies of large-scale production. Different components, features, accessories and finishes are then added to the basic design to produce a range of models to offer the customer.

- In the car industry the same chassis might be used as the basis for a family saloon, an estate car and a sports car with a powerful engine.
- Kitchen unit manufacturers produce a basic carcass onto which different arrangements and mixtures of laminates, doors, handles and drawers can be added.
- Textile manufacturers often offer the same pattern in different colourways and use a design for a range of furnishing products.

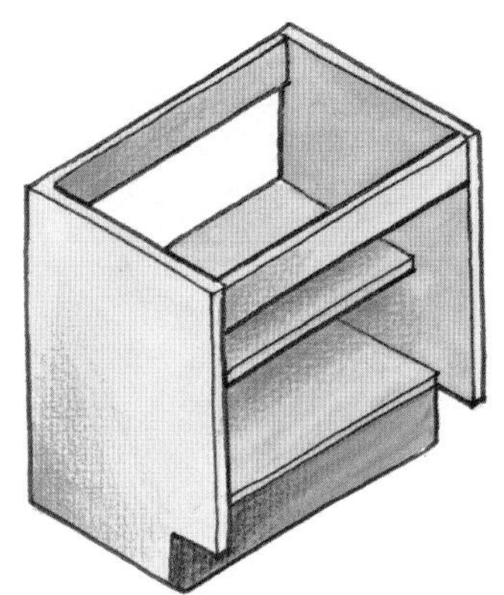

Any colour you like?

The techniques of modern mass-production were pioneered by Henry Ford in the USA between 1908 and 1916. He wanted to make a family car that a large number of people could afford.

Through the use of standardised parts, assembly and sub-assembly lines and an efficient organisation of labour, Ford reduced the time to put together his 'Model T' car from twelve and a half hours to two hours and forty minutes. Output and sales rose by over forty times, and prices dropped by a third.

However the cars were only available in black, because black paint dried quicker than any other colour.

Templates, Jigs and Moulds

Templates

One simple manufacturing technique is to use a **template**. This is where a standard shape is used as a pattern for cutting a series of identical shaped pieces. Another possibility is to cut or assemble a number of components all at the same time. Using a mould or a jig is another way of speeding up the production.

Jigs

In the manufacture of flat-pack furniture, a number of locating holes may need to be drilled in exactly the same place on a single component such as a upright. An effective method of doing this is to use a **jig**.

Here the machine operator only has to ensure the upright piece is located accurately each time and secured in the jig to guarantee accuracy in the drilling process.

Moulds

The use of a **mould** is very effective in the manufacture of door panels as used in kitchen design.

Companies such as Hygena vacuum form a plastic laminate and then glue this to the chipboard or MDF base. This ensures a consistent degree of accuracy.

Use CAD/CAM to help produce accurate templates and jigs to aid manufacture.

■ ACTIVITY

Design and make a simple jig that will enable you to drill a hole in exactly the same place on different pieces of wood.

IN YOUR PROJECT

- You might decide to use a particular material, or to develop a certain shape or form, simply because it will be easier to manufacture.
- How might a number of people work together to make the design more efficiently?
- Which stages of the manufacturing process might be automated?

Production Process Systems (2)

It is possible to define the step by step process by which most products are going to be made in simple terms. The different stages of manufacture can then be grouped together into key areas of production.

The **flow chart** shows the key stages in the production of a flat-pack kitchen unit.

ICT 

Some programs include symbols that can be used to quickly create and modify a flow chart.

When planning a production line, different operations can be coded by using different symbols.

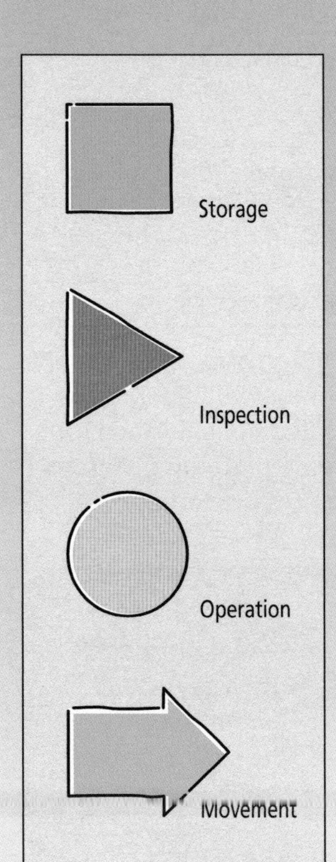

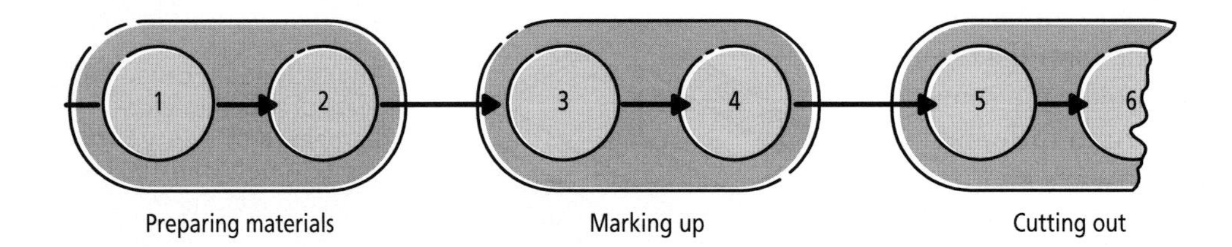

Flow chart grouped into simple stages

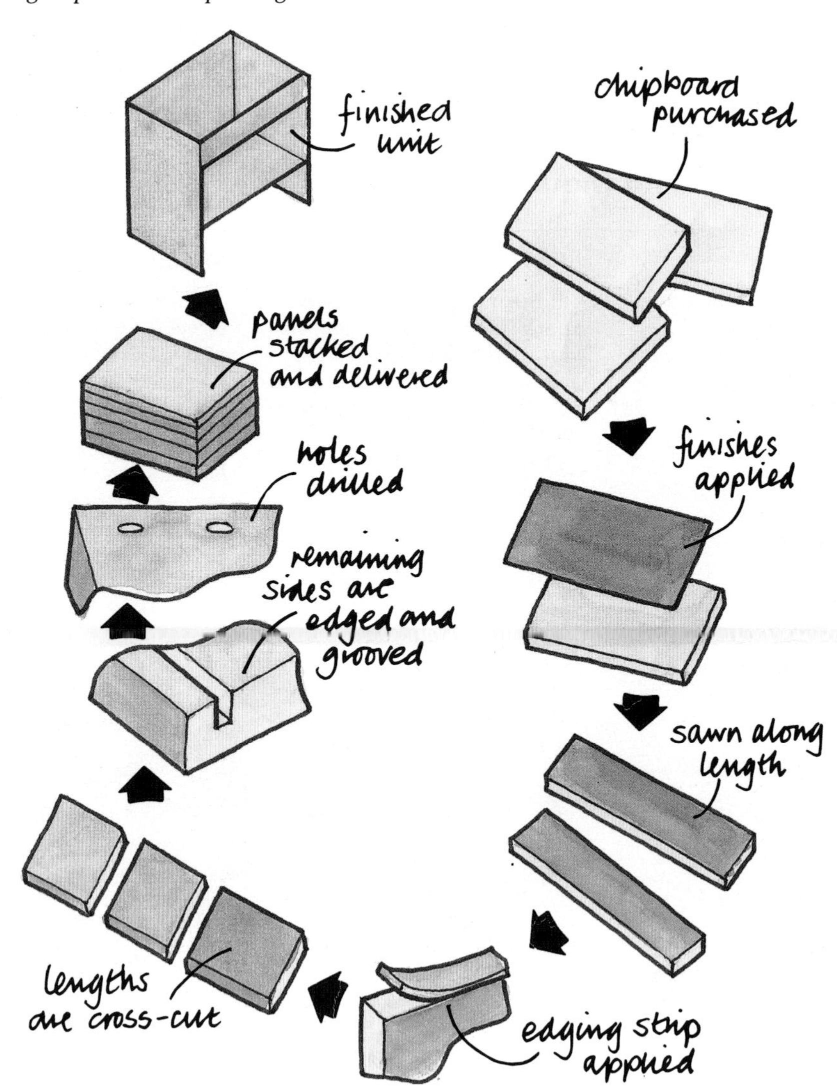

Sub-assemblies

When all the operations have been identified, the next stage is to plan the layout of the **production line**.

This is likely to involve a number of **sub-assemblies**, where groups of component parts are assembled before they are added into the main production line. In the manufacture of flat-pack furniture this may include the fitting of knock-down fittings.

Sub-assemblies are often made in what are called **manufacturing cells**. These are smaller individual units where four or five people operate specific machines or assemble items.

Other ways of organising a manufacturing process involve grouping similar machines and or materials together in one area. This has some advantages, but generally increases the distances that components need to travel.

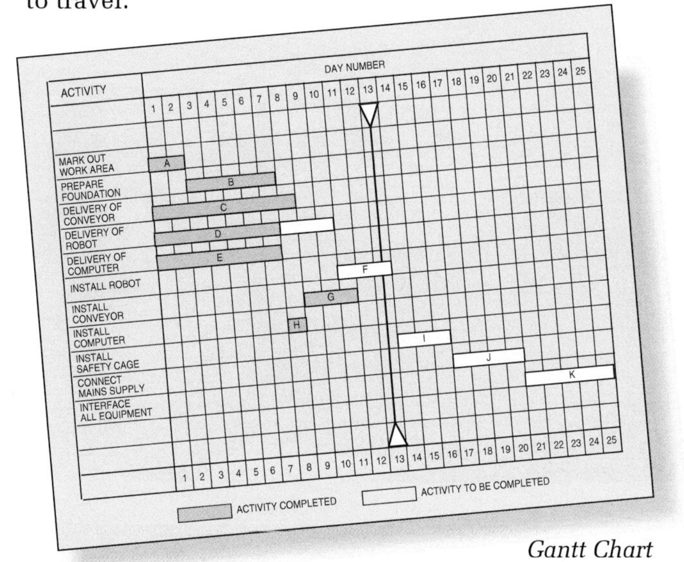

Gantt Chart

A conveyor-belt approach is well suited to some types of product, but relies on a constant supply of parts and continuous operation: a delay or breakdown at any stage can slow down or stop the entire system.

Just In Time

It is important to ensure that the correct materials and components arrive at the production line at exactly the right time and place. Many factories use a production control system called **just in time** to ensure this happens in the most efficient way.

Gantt Charts

In all types of production, a complex and accurate production schedule, such as a **Gantt chart**, is essential to tell everyone when to prepare, assemble and finish the different components.

Birds Eye Walls

Factory layouts and production scheduling can now be done efficiently using computer-based systems.

The Birds Eye Walls CAD system has the power to allow you to fly through 3D views of the factory and see how new layouts will look in practice. Production machines are built up on screen as a series of blocks. These can be quickly moved round to see where they fit best.

Layers (like sheets of tracing paper) are used within the CAD package to hold the details of different services such as water, steam, air and chemicals.

The system is much more accurate than old methods of drawing. Although the initial CAD drawing takes as long to produce, a great deal of time is saved because they don't need to redraw all the changes each time on a completely new drawing.

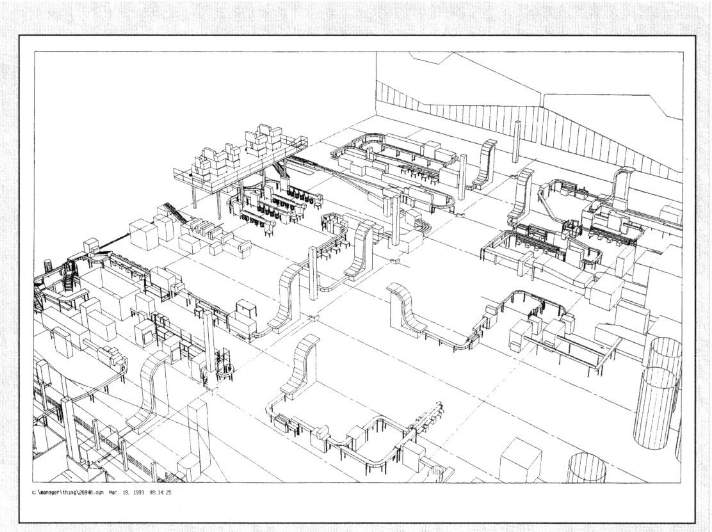

Manufacturing in Metal

Metals can be processed in many ways, including casting, rolling, forging, extrusion, machining, welding, bonding and anodising, to make a wide variety of products.

Case study:

Boosey and Hawkes

Musical instrument manufacturers Boosey and Hawkes were formed in the 1930s and make brass, woodwind and string instruments. They are particularly interested in the sounds metal tubes make when air is forced down them.

Metals need first to be extracted from their mineral ores. Sometimes they are combined to form **alloys**, such as steel (iron and carbon). They are then formed into bars, rods, sheets and tubes. These are then cast, deformed, machined and assembled into products.

Iron, steel, copper and aluminium are the four main metals used in industry.

Casting Processes

Casting involves pouring or injecting molten metal into a mould and then allowing it to cool. It solidifies into the shape of the mould. Complex, large or heavy shapes and a variety of surface finishes can be achieved with great accuracy.

There are a number of different methods of casting.

Process	Purpose	Tolerance (mm)	Weight of casting	Minimum section thickness	Labour costs	Equipment costs	Economic production levels	Typical products
Sand casting	Individual items	±1.5	0.5kg to many tonnes	2.5mm	Low	Low	Unit, small and medium batch production	Engine cylinder blocks, gearboxes etc.
Centrifugal casting	A hollow cylinder of uniform wall thickness	± 0.2	1kg to over 5 tonnes	2.00mm	Medium	Medium	Less than 1000	Pipes, lampposts, brake drums
Investment or 'lost wax' casting	Large volumes of high precision products	±0.05	Under 0.5Kg to over 100kg	4.00mm	High	Medium	100-5000	Small sized intricate components
Die casting	High volume production	±0.05	0.05kg to 50kg	0.5mm	Low	High	Minimum of 1000	Handles for DIY equipment, toy cars, casings etc.

The company aims to fuse traditional methods of manufacture with the innovation of modern technology. They use a variety of hand and computer controlled production processes and quality control systems to make products which need to both sound and look right.

Boosey and Hawkes use brass, silver and stainless steel which they cut, coil and crimp.

Computer numerically controlled (CNC) drills are used to obtain the fine tolerances needed for valves.

Finishing is very important. Buffing and polishing is essential to give the shine musicians expect.

Precision assembly and fine finishing are essential. Quality control plays a critical part in the manufacturing process.

Deforming Processes

Deforming means changing the shape of a piece of metal. There are a number of methods used to achieve this:

- ▷ rolling (squeezing the material between rollers)
- ▷ forging (using presses or hammers)
- ▷ extrusion (forcing the material to flow through a die).

These processes can be carried out at different temperatures, known as cold, warm or hot working.

To find out more about Boosey and Hawkes go to: **www.booseyandhawkes.co.uk**

ACTIVITY

See what further information you can find out about:

- ▶ how stainless steel, high speed steel, brass and aluminium are made
- ▶ the basic process of casting, deforming, cutting, finishing and joining.

Cutting Processes

There are a number of ways of **cutting** metals, such as

- ▷ **machining** (removing material with a cutting tool)
- ▷ thermal cutting (using gas, electric arc or laser)
- ▷ water-jet cutting (used for polymers and composite materials using robots).

Machining processes including simple hand sawing and filing, as well as drilling, lathe work and milling. Drilling machines cut holes. The lathe is used for producing cylindrical components. A milling machine produces slots or shapes in the surface of materials.

Finishing Processes

The main methods of **finishing** metals include:

- ▷ shot blasting (an abrasive cleaning process)
- ▷ pickling (removing scale after components have been hot worked)
- ▷ tumbling (for removing rough edges)
- ▷ electroplating (to improve appearance)
- ▷ anodising (to provide a corrosion-resistant surface)
- ▷ painting (for corrosion protection and an attractive appearance).

Joining Processes

There are five main ways of **joining** metals together:

- ▷ mechanical fastenings, such as screws, nuts and bolts which can be semi-permanent or permanent
- ▷ soldering
- ▷ brazing
- ▷ welding
- ▷ adhesive bonding.

Manufacturing in Plastics

Moulded plastic products require little or no finishing processes. The main processes are:

- ***extrusion***
- ***blow moulding***
- ***vacuum forming (see page 34)***
- ***injection moulding***
- ***forming using glass reinforced polyester (GRP)***
- ***coating***

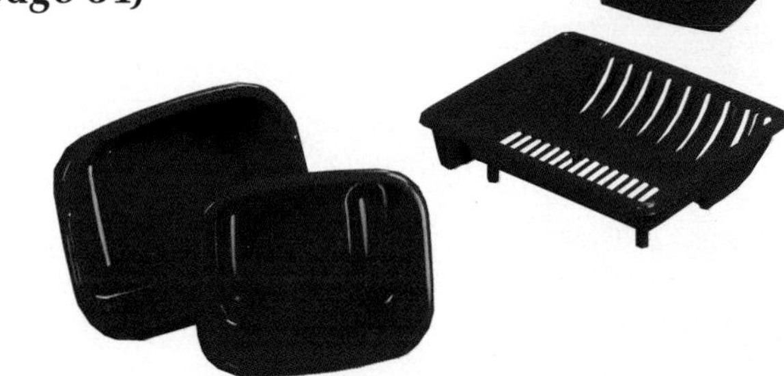

Extrusion

Thermoplastic material in powder or granule form is fed into the machine and heated. When soft it is squeezed through a die which determines the profile of the extrusion. When the extrusion has cooled and hardened it is cut into the required lengths.

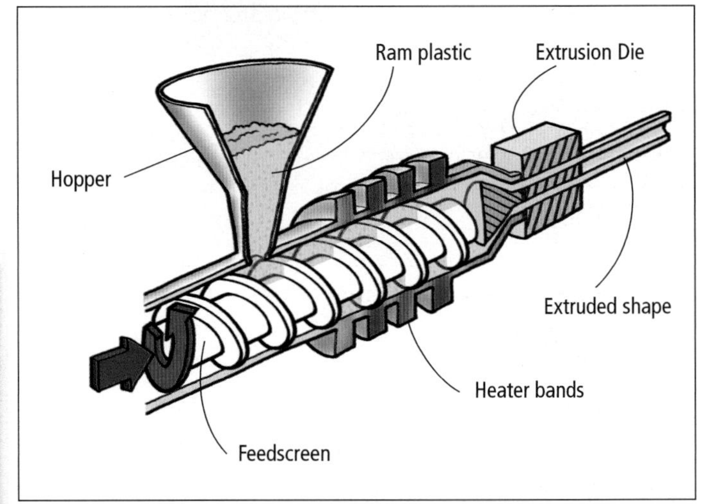

Typical products made by **extrusion** include tubes, rods, and pipes. A wide range of sizes and shapes are possible. It is a fast process which is suitable for high volume production.

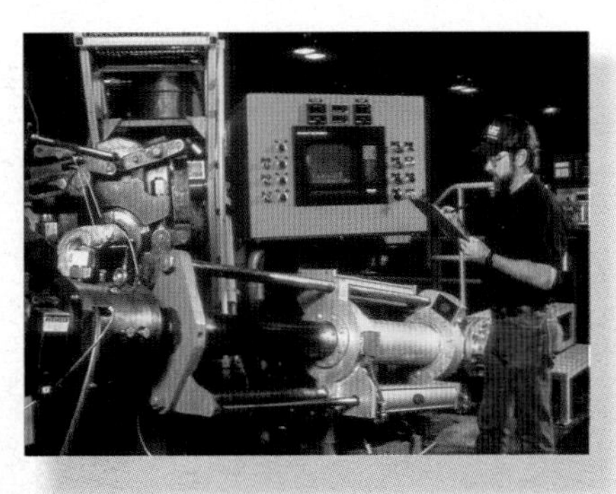

Blow Moulding

The material used for **blow moulding** is thermoplastic sheet, the same as that used in vacuum forming. Warm compressed air is forced against the softened sheet, causing it to expand either into a dome or against a mould, taking up this shape as it cools.

Blow moulding is used to produce hollow products such as plastic bottles for drinks, detergents, cosmetics etc. It comes in a multitude of shapes, sizes and colours.

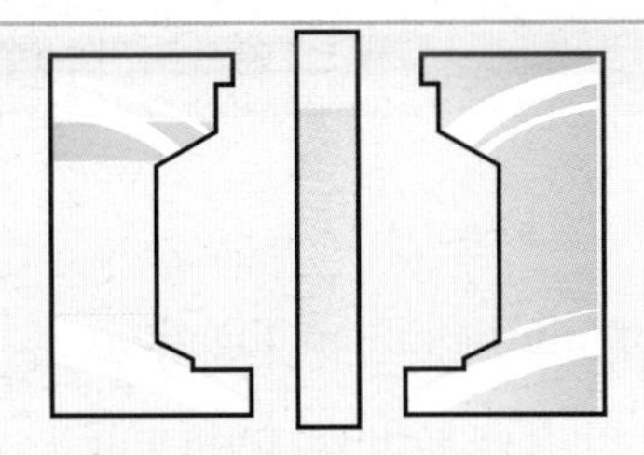

1 Thermoplastic tube is extruded between two mirror image halves of a mould.

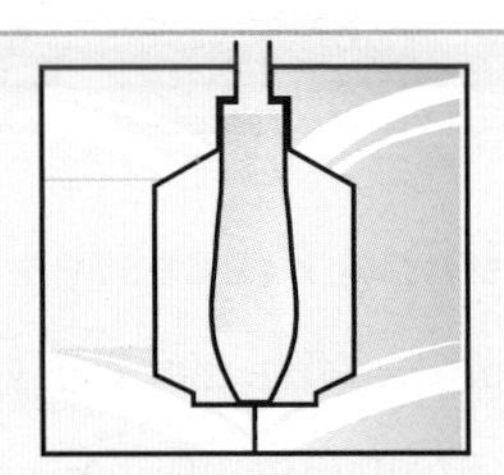

2 The mould is closed, sealing the bottom of the tube and at the same time cutting it to the required length. At this stage the piece of tube is called a parison.

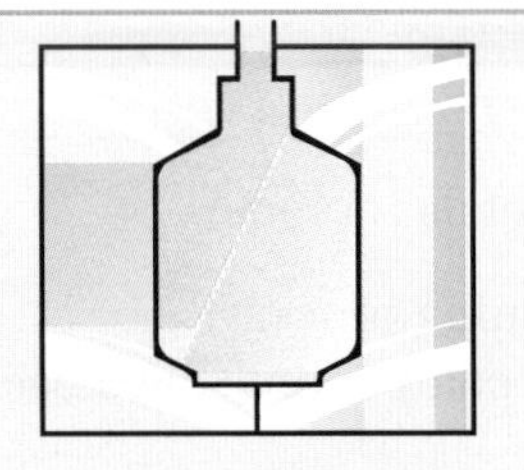

3 Warm air is then blown into the softened parison, forming it against the walls of the mould.

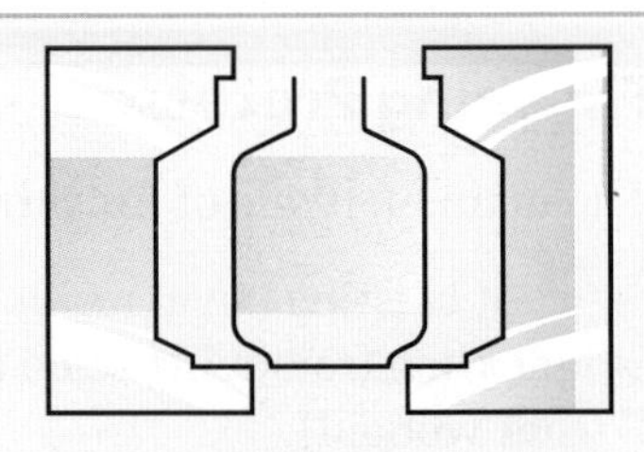

4 On contact with the cold walls of the mould the plastic cools, the mould is opened and the now cold, blown shape falls out.

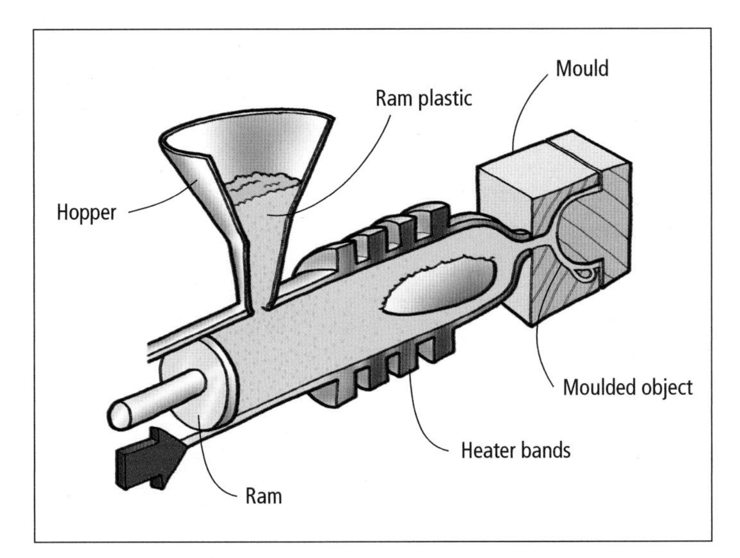

Injection Moulding

Thermoplastic powders or granules are heated to a fluid state and then injected into a metal mould where they cool and harden.

Injection moulding is used to produce very large production runs of things like polypropylene chairs, washing up bowls, buckets, etc. It provides an excellent surface finish.

Glass Reinforced Polyester (GRP)

Layers of fibre matting are placed over a former, and a liquid resin poured over. The resin hardens to produce a rigid material. Further layers can be added. **GRP (glass reinforced polyester)** is used for larger and more complex shapes such as boats.

GRP has many advantages over other materials. It is:

- strong – double the strength/weight ratio of steel
- resistant to weather, oil and water
- easy to form
- very cheap.

Fibreglass fumes or particles do however pose a serious health hazard. Protective clothing and breathing filters must be worn.

Coating

Plastic coatings provide a waterproof protection for other materials. Dish drainers, vegetable racks, coat hooks, door handles are all finished in this way. There are several types of coating which can be used:

- cold dip coating (for very thin coatings)
- hot dip coating (for thicker coatings)
- fluidized dip coating (sticks to metal)
- spraying (for large items).

www.

To find out more about plastics go to:
www.materials-database.org.uk
To find out more about TVR go to:
www.tvr.co.uk

Case study:

TVR car body mould

Unlike most car bodies, the ones made by TVR at their Blackpool factory are made from rolls of glass fibre sheet.

The sheets are placed into moulds and liquid resin is added. After about 20 minutes the resin sets hard, and another layer can be put on top.

When the mould is finally removed the glass-fibre shell is revealed. The body needs finishing and painting.

Production is much slower than a large manufacturer – the company only makes 22 cars a week. Tooling costs are considerably less, however, and the finished car is strong, light-weight and fast.

TVR place great value on traditional materials such as leather, wood and aluminium.

Manufacturing in Wood: Case Study

Christies is a luxury fitted furniture company. They specialise in producing one-off, made-to-measure bedrooms, bathrooms, home offices and kitchens. By using a sophisticated CAD-CAM system they are able to design, make and deliver furniture and fittings individually for each customer within two weeks.

Made to Measure

Each order Christies receive is made to measure. There is no pre-cut stock because they do not make the furniture until the exact measurements of the customer's room are known. Odd-shaped rooms and sloping ceilings provide no difficulty for their CAD system, which automatically calculates all the complex angles, shapes and corners needed.

Customers can choose from an extensive series of ranges and colour and style combinations. Christies' designers combine their furniture with co-ordinated sanitary ware, tiles, lighting, carpets, mirrors and other accessories. They also include the latest in bathroom control systems technology which fills the bath to a pre-programmed depth and temperature, stops the water running and then bleeps when it's ready!

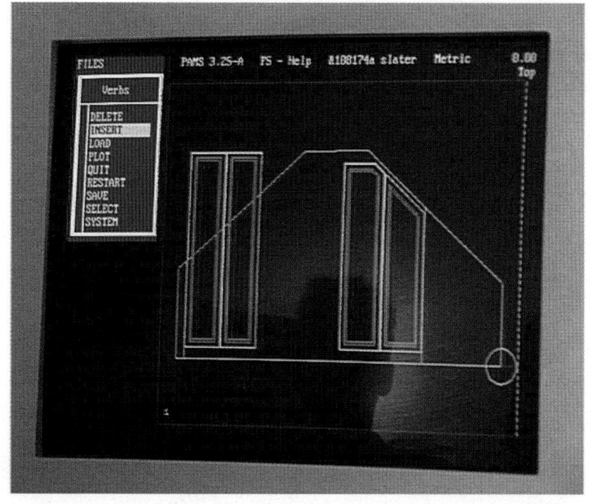

The company pioneered the use of **high density fibreboard**. **HDF** can be machined, drawn, drilled and cut to a very high specification. It also provides an excellent surface for paint, varnish and lacquer finishes.

Environmental concerns

HDF is made from young saplings that would otherwise be discarded. In the factory a dust extraction system collects the sawdust from the machines. This dust is then burned in a special incinerator to provide the heat for their factory and offices.

Design to Delivery

The process begins with a site visit to measure up. This is then reproduced in 3D on a CAD system, so that the client can see what the designer proposes and discuss styles and colours.

Once the design has been finalised the order is despatched to the factory floor.

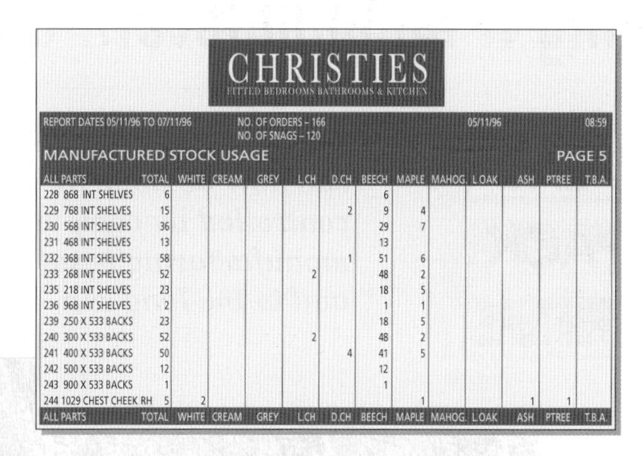

CHRISTIES
FITTED BEDROOMS BATHROOMS & KITCHEN

REPORT DATES 05/11/96 TO 07/11/96 | NO. OF ORDERS – 166 | NO. OF SNAGS – 120 | 05/11/96 | 08:59

MANUFACTURED STOCK USAGE | PAGE 5

ALL PARTS	TOTAL	WHITE	CREAM	GREY	L.CH	D.CH	BEECH	MAPLE	MAHOG.	L.OAK	ASH	PTREE	T.B.A.
228 868 INT SHELVES	6						6						
229 768 INT SHELVES	15					2	9	4					
230 568 INT SHELVES	36						29	7					
231 468 INT SHELVES	13						13						
232 368 INT SHELVES	58					1	51	6					
233 268 INT SHELVES	52				2		48	2					
235 218 INT SHELVES	23						18	5					
236 968 INT SHELVES	2						1	1					
239 250 X 533 BACKS	23						18	5					
240 300 X 533 BACKS	52				2		48	2					
241 400 X 533 BACKS	50					4	41	5					
242 500 X 533 BACKS	12						12						
243 900 X 533 BACKS	1						1						
244 1029 CHEST CHEEK RH	5	2						1			1	1	
ALL PARTS	TOTAL	WHITE	CREAM	GREY	L.CH	D.CH	BEECH	MAPLE	MAHOG.	L.OAK	ASH	PTREE	T.B.A.

Organising and monitoring production is crucial to keep delivery times to the minimum.

HDF and chipboard panels are taken from storage and sawn to shape before being drilled, edged and routed using computer numerically controlled (CNC) machines (see page 136), directly supplied with data from the computer design.

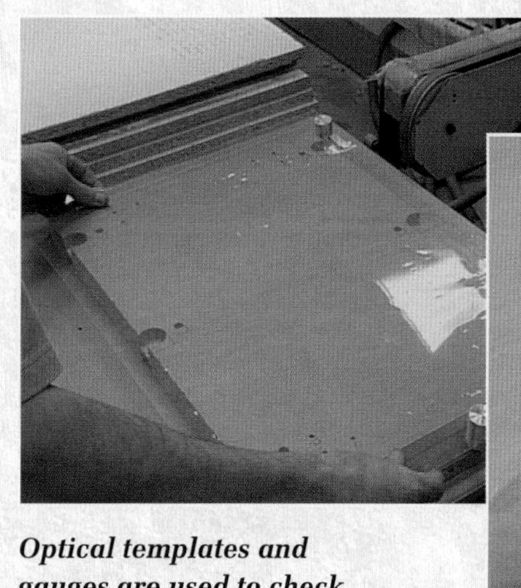

Optical templates and gauges are used to check positions and depths of holes for fittings.

A CNC (computer numerically controlled) routing machine automatically produces the relief patterns on cupboard doors.

The panels are then sanded, painted and finished before a final quality control check.

Making it by Computer

An increasing number of production processes can now be done by machines which are controlled by computers. Automated manufacturing is safer, quicker, more reliable, and in the long run, cheaper.

Computer-aided manufacture (CAM)

Computer-aided manufacture (CAM) is a term used to describe the process whereby parts of a product are manufactured by equipment which is controlled by a computer.

One of the restrictions of batch-production is that after a relatively small number of products have been made a machine has to be re-set to the requirements of a different product.

The main advantage of CAM is that the new instructions are stored electronically and can be down-loaded and programmed into the machine very quickly. This also facilitates making small changes to the design to suit changes in the market or to produce specialised short-run products for individual clients.

Where computer-aided manufacture is used to replace a manual operation, greater productivity is possible, because the machine can work continuously. There is also a greater consistency of quality, and fewer faulty goods. CAM systems can also work with materials and chemicals which might be harmful to human operators.

In many cases the manufacture of a complete product is a combination of CAM and hand operations.

Industrial robots can be programmed to perform a variety of tasks using a range of tools and materials.

An example of this may be where the cutting and machining of a panel is completed using the CAM facility but the application of a suitable finish is done manually.

This is particularly effective where a number of smaller manufacturers specialise in the making of a component which contributes to a whole product. The car industry relies on this practice a great deal.

Automated and autonomous guided vehicles can be used to transport components, tools and materials to the appropriate machine.

Computer numerical control (CNC)

Computer numerically controlled (CNC) machine tools can be independently programmed, but also have the facility to exchange data with other computers. They therefore become part of a complex automated production system.

Computer integrated manufacture (CIM)

CAD and CAM systems contribute to the development of **computer integrated manufacture (CIM)** This is a totally automated production process with every aspect of manufacture controlled by computer.

Powerful CAD systems can be linked into a CIM system. This allows the entire design development, production

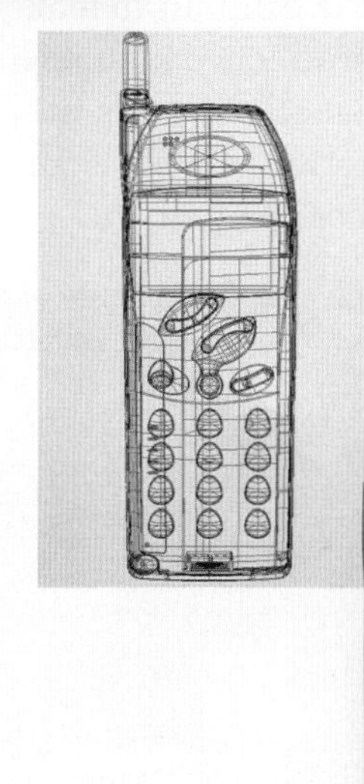

schedule and manufacturing operation to be undertaken by a single system.

Manufacturing companies which have adopted such systems have been able to make dramatic reductions in their prices, and increase their quality and reliability.

A virtual prototype

A **virtual prototype** is a very high quality computer-generated image of a product, viewed on screen or printed out, which provides a photo-realistic impression of what it will look like when made. These are used to help communicate a design idea to the client in a convincing way. The great advantage over the traditional airbrush is that it is easy to make detail changes without the need to recreate the entire image.

Making it somewhere else

New ways of creating and sending are being developed for sending information electronically. Computer-generated data about a final design can be sent almost anywhere in the world in a few seconds. The data is then fed directly from the computer into manufacturing equipment to make the product.

CAD/CAM case study:

The AKA mobile phone

The first stage was to create two appearance proposals for the design of this new portable telephone. These were presented as hand-made models. The models were made from a material called Ureol which is a resin that can be cut and finished easily by hand.

Next AKA were commissioned to combine the designs and the technical specification into a viable product. Alias software was used to create digital models of all the electronic components and to then wrap a 3D 'skin' round them. Buttons were positioned to ensure existing components could be used, saving development time and costs.

Shaded images of the wireframe were constantly assessed to ensure that the model met the aesthetic as well as technical requirements.

AKA then created a physical model to present to the client. This one was made directly from the computer data using a CNC milling machine.

A **rapid prototype** was produced to check that all the electronic components and plastic housing fitted correctly. This was created using a laser cutting a liquid resin, again fully computer controlled from the original computer data.

Fully rendered colour images were then created. Colours, textures and graphics were assigned to the surfaces of the digital model and photo-realistic images created. These pictures were used in brochures and advertising posters for the product launch.

Finally the 3D digital data were sent via a modem directly to a toolmaker in Seoul, Korea, where the production tools were made.

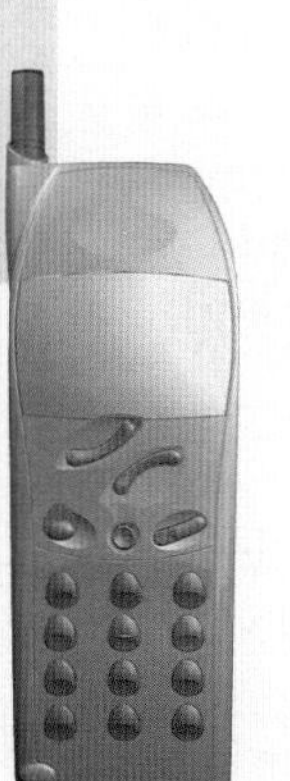

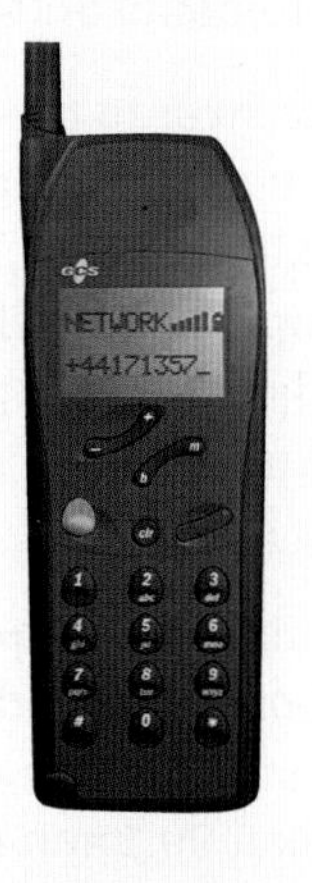

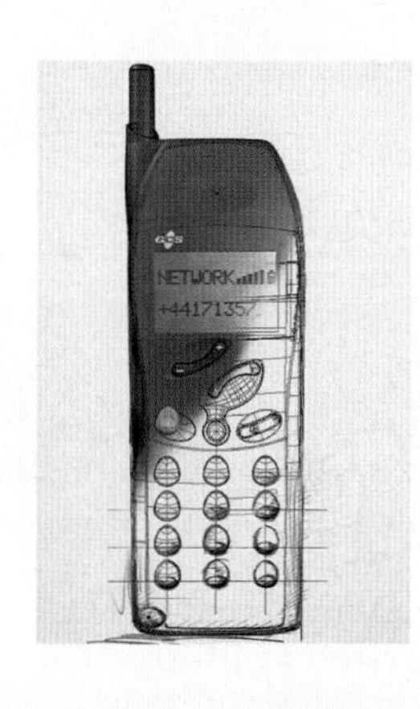

Tapping into a new dimension

3D scanning has been developed to make computer reproductions of solid objects, so that designers can model the shapes they want in more traditional, human-friendly ways such as carving wood or shaping clay rather than drawing 3D shapes on a 2D screen.

A research team have developed a method of scanning objects with a laser, and transmitting the data by phone to a company that can create plastic models from computer instructions.

This process means that one day you will be able to down-load a reproduction of an antique bronze or a product designed on the other side of the world.

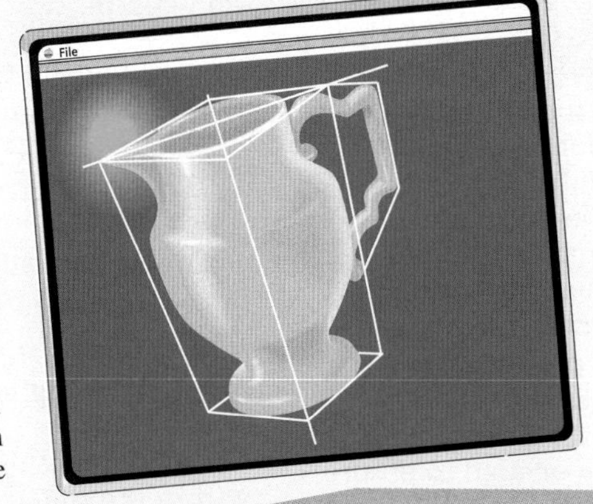

Based on an article by Chris Partridge in 'The Times', 24 July 1996.

Quality Counts

Manufacturers need to ensure that all the products they are producing are of an acceptable standard. A range of techniques has been developed to help check and maintain quality over a long production run.

Use CAD to produce accurate automatically dimensioned formal drawings.

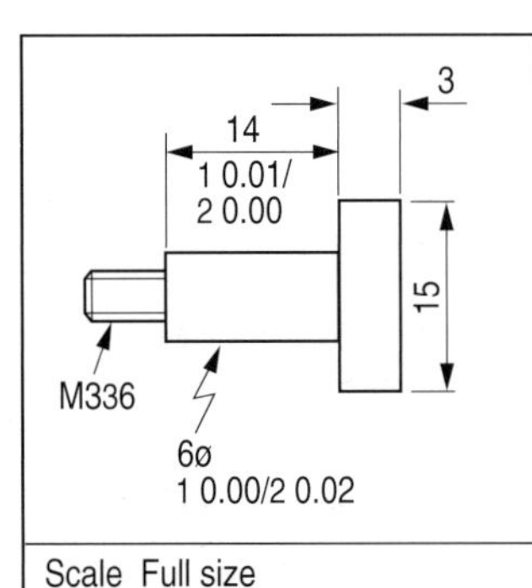

IN YOUR PROJECT

- How accurately do the different parts of your design need to be made?
- Which components need to fit together most accurately?

KEY POINTS

- The specification for a product should include a statement about its tolerance limits.
- Tolerances are important to ensure reliability, which in turn reduces wastage of products during manufacture.
- Testing procedures are needed to ensure items are within the stated tolerance limits.

Working to a Tolerance

As you have probably discovered it is difficult to make something accurately. In complex products a high degree of accuracy is essential to ensure that all the parts fit together exactly. The important question is how accurate does it need to be? This will vary according to the requirements of the product.

The answer to this question is known as the **tolerance level** – the extent to which the size of a component must be accurate. This is usually expressed by two numbers: an upper and lower limit.

In a simple example, a component intended to be 100 mm in length could vary between 99.1 mm and 100.9 mm. The tolerance is the difference between the upper and lower limits, i.e. 1.8 mm or + or –9 mm.

New automated equipment tends to be quicker and more efficient at producing and testing components which are finely toleranced.

In the manufacture of furniture, the higher the degree of accuracy, the better the quality of the product. Achieving this greater accuracy requires careful measurement and skill in controlling tools. This usually means that the cost will be higher.

Testing procedures are needed to ensure items are within the stated tolerance levels. These are usually set by the manufacturer, but in some cases they will be set by the client.

ACTIVITY

Try to cut three pieces of material to exactly the same length. Check to see if they are the same. Set your own tolerance level. Can you suggest an aid which when used with the operation will improve the accuracy of cutting?

TOLERANCE LIMIT
24.90
ACTION LIMIT
WARNING LIMIT
24.94
SAMPLE MEAN SIZES PLOTTED
SAMPLE MEAN SIZE
WARNING LIMIT
ACTION LIMIT
TOLERANCE LIMIT
24.98

Quality Control Systems

One way to check the quality of products as they are made is to check every one to ensure that each is satisfactory. A more sophisticated and quicker approach is to set up a system of **quality control**. This involves inspecting a sample of components as they are made, and the gathering and analysing of records of the samples.

In simple terms, a sample of components (say 1 in every 100) are subjected to rigorous tests which identify and record how close the item is to its target. Provided it is found to be within acceptable limits, production continues.

By examining the pattern of a series of tests it may be possible to notice that a particular machine is increasingly producing components which are getting close to unacceptable tolerance limits.

In such a situation it is possible to adjust, or if necessary repair, the machine before it starts to produce items which would have to be classed as defective and possibly shut down the production line. The aim of quality control is therefore to achieve zero defects by predicting the failure of a machine before it happens.

The use of automated testing machines and of electronic gathering and analysis of data leads to higher standards of quality and less wastage.

IN YOUR PROJECT

- Identify when the best time is to check for accuracy.
- What type of testing is to be carried out?
- How often do you need to complete these checks?
- Which British Standards would apply to the production of your design?

KEY POINTS

- Quality control systems help manufacturers reduce wastage and delay in production.
- They do this by predicting failure or other potential problems before they happen.
- Quality control production checks form a specific part of a broader programme of quality assurance across a company.

Quality Assurance

Quality assurance is the overall approach that ensures high standards of quality throughout the company. It includes the development and monitoring of standards, procedures, documentation and communications across the company as a whole. Usually a quality manual is produced which contains all the relevant information to guide staff.

British Standards

One of the major bodies which promotes quality is the British Standards Institute. They produce documents that clarify the essential technical requirements for a product, material or process to be fit for its purpose. There are over 10,000 **British Standards** for almost every industry from food to building construction and textiles to children's toys.

Certification that a product conforms to a stated British or European standard provides an assurance that an acceptable quality can be expected. This reduces the risk of someone buying a product which could be defective in some way.

If you want to find out more about the British Standards Institute, go to: **www.bsi-global.com**

Standards Catalogue
1996
1997
BSI

Designing for Manufacture

There are many different things that need to be considered when designing something suitable for manufacture. These include its life-expectancy, maintenance, and the need to reduce costs.

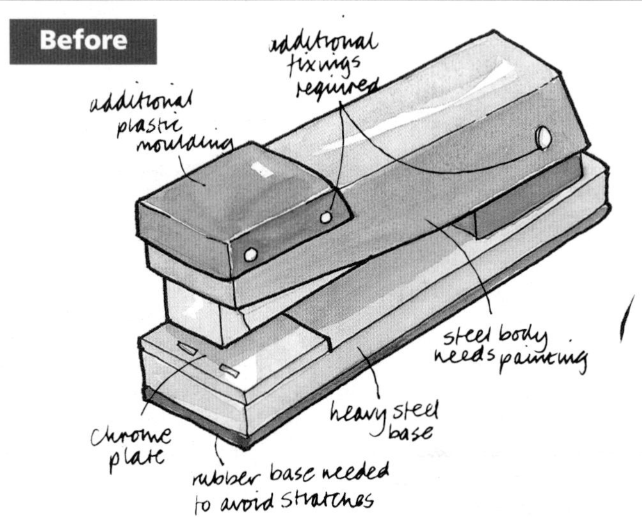

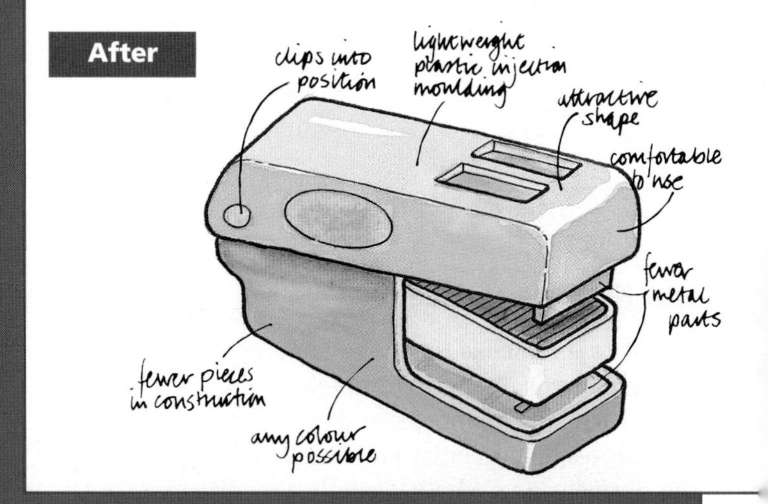

Cost, Durability and Maintenance

Counting the costs

When a manufacturer is faced with a new product a major consideration has to be how to keep **production costs** to a minimum. Some elements will be **fixed costs** while others are known as **variable costs**.

Fixed costs are those incurred in setting up the production line such as machines, the tooling and factory space.

Variable costs are likely to be the cost of materials, energy used, the wages of the workforce, insurance and maintenance, etc.

The costs of storage, transportation, packaging and selling all need to be taken into account as well.

The actual cost of the manufacturing of a product in terms of its materials and labour will vary but usually only represents some 5–10% of the final selling price.

Design for life expectancy

Many examples of early pre-packed, DIY furniture, had design faults or were poorly manufactured or used inferior materials and fittings.

Customers expect a certain minimum time that a product such as a drawer unit will work and a product that fails before a reasonable time could be very costly for the manufacturer to repair, or in some cases replace.

The main problems with furniture that has been produced using manufactured boards tends to centre on the following:

- the chipboard fractures where there are weaknesses in the design.
- the fixtures or fittings such as handles or drawer-runners break.
- the lipping comes away from the edge
- certain fittings are missing.

If the product suffers from any of these then the customer is unlikely to make a repeat purchase of the same brand and the brand might develop a reputation for being unreliable.

However, if the product lasts for many years consumers will not need to buy replacements so often and demand will fall. The number of products which need to be manufactured will therefore drop and as a result the price may well rise. Many products therefore contain components that are likely to fail after a number of years and which would be very expensive to replace. This is called **planned obsolescence**.

Design for maintenance

Designers have to consider how often a new product will need to be maintained during its usage, and take this into account while developing ideas. They will also have to think about how easy it will be to undertake the maintenance work. If a component needs cleaning, adjusting or replacing by the user (e.g. replacing a battery) it must be quick and easy to do. Other maintenance however might need to be done by trained specialists and providing easy access might result in damage if the user tries to do it themselves.

Ideally a product should be maintenance free, but this is likely to involve the use of higher quality components and finer tolerances in manufacture, which will inevitably increase the cost.

"We needed a design that was easier, and cost effective to manufacture in large numbers, used fewer parts, was attractive and user-friendly"

IN YOUR PROJECT

Make sure there are statements in your final production design report about:

- how much it will cost to make in quantity
- how long it might reasonably be expected to last without breaking or wearing out
- how frequently and easily different parts will need to be maintained
- aspects of the production which might be hazardous
- what steps you would recommend to be taken to minimise the risks, and why
- how moral, social and cultural issues have been considered.

Hazards and Risk

Health and safety

The world is a dangerous place. As designers and manufacturers produce new products, they need to ensure that they will be safe to use and also safe to make.

Safe to use

The designer must ensure that the product conforms to all the relevant safety standards, including those of other countries in which it might be sold.

Careful consideration must be given to ways in which people might misuse the product, and any necessary safety devices and warning labels included: the designer can be held responsible for any accidents which occur as a result of poor design.

Safe to make

There are four main areas to consider to help avoid potential accidents:

- the design of machinery and tools being used in the manufacturing process
- the physical layout of the work area
- the training of the workforce
- the safety devices and procedures.

In the manufacturing process there are a series of regulations and codes of practice which must be observed.

It is also essential to reduce the number of potential hazards – unsafe acts or conditions – which could occur in the workplace. Accidents are extremely costly in terms of personal distress, compensation and lost production.

Reducing the risks

Although we cannot avoid taking risks, we can take steps to assess the likelihood of something happening, and minimise its impact if it does.

As well as the legal requirements and more general codes of practice for health and safety, a considerable amount of documented information is available to help guide the design of safe products and working environments.

Ergonomic studies and **anthropometric data** can be used to determine optimum positions for displays and controls on products and machines, and the most suitable sizes and arrangements for work-spaces and conditions (e.g. the distribution of light; noise; heating and ventilation).

Risk assessment

When a production process involves hazardous situations it is necessary to analyse and assess each particular risk situation and ensure that adequate precautions are taken to minimise the potential danger.

It is the responsibility of an employer to assess the risks involved in each stage of production and justify the level of precautions adopted to a **Health and Safety Inspector**.

Design Issues

When specifying the requirements for a mass-produced product, designers need to consider a wide range of moral, economic, social, cultural and environmental issues. These often produce conflicts which can be hard to resolve.

Moral issues

In certain situations a product may have the capacity to injure or harm someone – either the user or a bystander. Cigarettes and alcohol are obvious examples. Bull-bars on cars may look good and help improve sales, but they are likely to increase the severity of injury to a pedestrian in an accident.

Social issues

Some products can have a major impact on the way in which large groups of people live their lives. Convenience foods, for example, mean that there is less likelihood of the family sitting down together to eat a meal. Promotion and packaging can help counter this by providing two-person portions and using images of family meals.

Information and communication technologies are in the process of making a major impact on society, as work and shopping can be increasingly undertaken at home. Advanced automation reduces the number of people needed to produce and distribute goods, causing unemployment.

Cultural issues

The particular beliefs, ways of life and traditions of different groups of people have a major effect on the way they live their lives – what they do, where they go, and the things they buy. Food and clothing and the symbolism of certain shapes and colours all play highly significant roles in maintaining the identity of a particular culture: when a product is intended for use by a range of cultures it is important to identify and recognise such needs.

Final Presentation, Testing and Evaluation

At the end of the project you should have:

- ***a model of the design which represents what it would look like if it were mass-produced***
- ***coloured drawings to show clients and potential users the main features of the design***
- ***workshop drawings which could be used by people making the product in a factory***
- ***samples of the instruction leaflet which will tell users how to put the product together***
- ***a short report on how your design would be manufactured.***

Use a Graphics and/or DTP program to lay out text and artwork.
Use a Presentation package to help communicate the features of your final design proposal.

Self-assembly Instructions

Have you ever tried to assemble an item of furniture from an instruction sheet? This is often difficult because:

- the written instructions are not clear enough
- the drawings don't look like the actual components
- the sequence of assembly is not always correct
- extra tools (e.g. a specific type of screwdriver) are needed

What instructions would be needed for someone to make your product? Plan and make a simple leaflet to help explain the process of assembly.

- Work out the order of construction.
- Write simple instructions, referring to illustrations.
- Use a variety of drawing techniques, e.g. plans, elevations, isometrics, cross-sections, cut-away views, etc. to make things as clear as possible.
- Use colour and texture to code and identify materials and components.
- Use arrows which express the direction and amount of force needed to turn and fix things together.

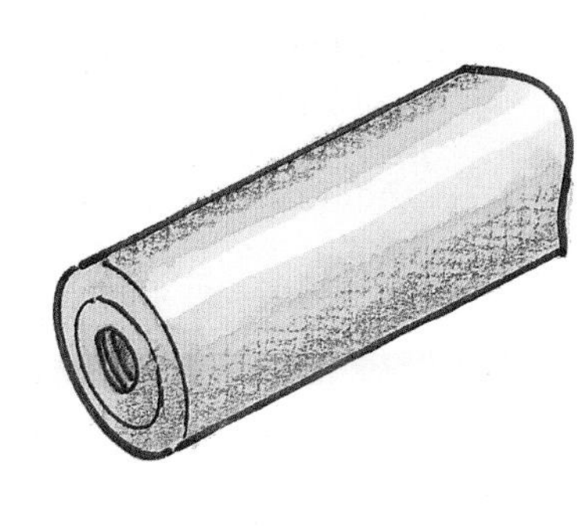

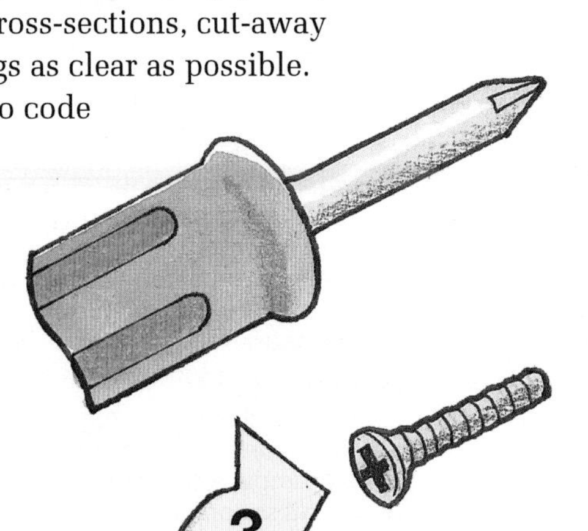

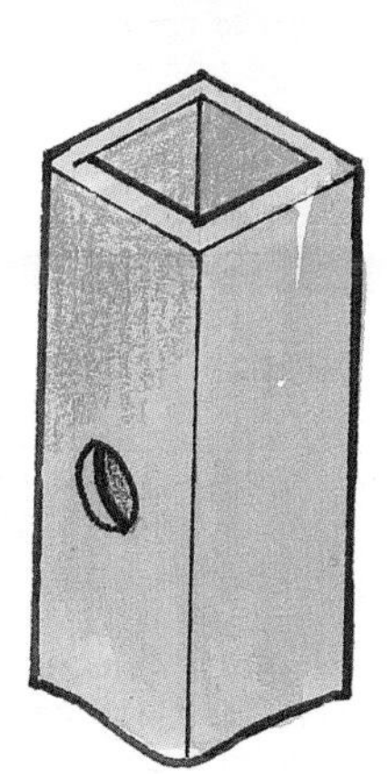

Testing the Final Product

If your product is aimed at a specific range of users then it is they who should be the focus of the testing.

You will need to record their views on the product: what they like and dislike about it, and why. Don't be too disappointed if your design is criticised at this stage. Respond positively to the comments made and feel reassured that the product would be even more successful if you had the opportunity to develop the design further.

You could also devise some tests to measure general and specific things such as:

- the ease with which people could assemble the product.
- if there is any deterioration of ease over 100 pulls of a drawer or other moving part.

Your tests should be based on the requirements of your design specification.

Testing the Final Presentation

You could also assess how well people have understood your design proposals from your models, drawings and presentation work. For example you might ask:

- what sort of people is the design aimed at?
- what words would you use to describe its colours, textures and shapes?
- what are its particular design features which make it different from other similar products already on the market?

You could test the effectiveness of your assembly instructions by seeing if someone can construct your design without you being there to guide them.

If you have had to change your original design to make it more suitable for batch or mass production you may need to produce a revised series of drawings and/or models.

Final Evaluation

In your evaluation describe the tests you have carried out, perhaps recording the results in chart form. Discuss the quality of the product (including how well it meets the specification) and the processes you went through while designing it.

In particular refer to how your final design needed to change to take account of the demands of mass-production.

Try to give a balanced evaluation weighing up the strengths against the weaknesses of your work, and your design.

If you had more time to develop your design further what would you do?

Examination Questions

You should spend about one and a half hours answering the following questions. To complete the paper you will need some A4 and plain A3 paper, basic drawing equipment, and colouring materials. You are reminded of the need for good English and clear presentation in your answers.

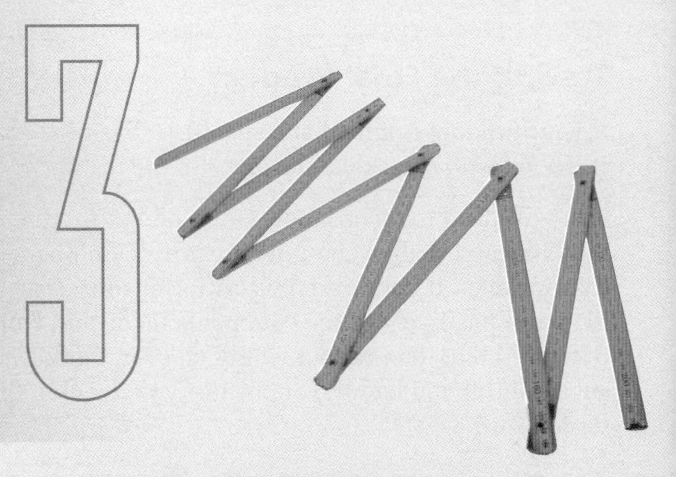

1. This question is about design specification. See pages 16-17. *(Total 12 marks)*

You have been asked to design a product with the following specification:

- a low cost, flat-pack item of furniture which will be used in a hall to store coats, shoes, hats and pairs of gloves for a family of four.
- enough space for each member to have storage for one of each item of clothing, but be small enough to fit into a hall 2 metres x 2.5 metres and still leave space for people to move around.

List six things that the hall storage unit will have to do. *(12 marks)*

2. This question is about product safety issues. See pages 92-93 and 141. *(Total 2 marks)*

Name two safety issues that you will need to think about when designing a piece of furniture that will be used by children and adults.

(2 marks)

3. This question is about designing. See pages 18-19. *(Total 20 marks)*

a) Sketch two ideas for a hall storage unit. Your ideas need to match the design specification in question 1. *(6 marks)*

b) Choose one idea and develop it. You need to include information about materials, finishes and colours. *(3 marks)*

c) Look back at question 2 for your safety issues. Make notes on your drawing about how to make it safe to use. *(2 marks)*

Marks will also be given for:

d) communicating your ideas by using your drawing skills. *(6 marks)*

e) Explaining your ideas by using appropriate technical words. *(3 marks)*

4. This question is about the characteristics of materials. See pages 110-111 and 32-33, 54-55 and 82-83. *(Total 18 marks)*

a) Name three materials that may have been included in the hall storage unit shown below. *(6 marks)*

b) Say which part each material has been used for. Give one reason for this choice. *(12 marks)*

Copy and complete the table below to answer this question.

Material	Part used for	Reason for choice
1. 2. 3.		

5. This question is about industrial practice: types of production. See page 126. *(Total 4 marks)*

a) Which type of production would be most suitable for making 10,000 hall storage units? *(2 marks)*

b) Which type of production would be most suitable for making milk bottles? *(2 marks)*

6. This question is about components. See pages 110 and 120-121. *(Total 10 marks)*

Look at the illustration of the hall storage unit again. The manufacturers want to use fixings from outside suppliers.

a) Name two ready-made components which can be used to make assembly of the unit quick and easy. *(4 marks)*

b) Explain where you would use these on the unit. *(4 marks)*

c) Why do push-fit connections have economic advantages? *(2 marks)*

7. This question is about industrial practice: tolerances. See pages 138-139. *(Total 6 marks)*

The diameter of the hole for the light fitting in the top of the hall storage unit on page 102, needs to be 100 mm + or – 0.5 mm.

a) What does this mean? *(2 marks)*

b) Using notes and sketches, show how this can be quickly and easily checked. *(4 marks)*

8. This question is about using ICT in designing and making. See pages 136-137. *(Total 8 marks)*

a) What does CAM mean? *(2 marks)*

b) Describe two ways in which CAM could be used in the manufacture of the hall storage unit shown below. *(6 marks)*

Front elevation **Side elevation**

vertical shelf support

backing board

glass shelf

horizontal shelf support

9. This question is about production process systems. See pages 110 and 128-129. *(Total 10 marks)*

Look at the photograph of the three wall mounted shelving units shown below on the left and their front and side elevations.

a) Write down, in the correct order, the main processes in the making of the backing board of one of these products. *(5 marks)*

b) Now do the same for the glass and aluminium tube shelves. *(5 marks)*

10. This question is about manufacturing safety issues. See pages 140-141. *(Total 4 marks)*

What are the four main areas to consider to help avoid accidents in the workplace? *(4 marks)*

11. This question is about evaluating the final product. See pages 142-143. *(Total 6 marks)*

Give details of two ways of testing the success of a hall storage unit. *(6 marks)*

Total marks = 100

Computer at Work

Ergonomics (pages 80 and 114)

Designing for Manufacture (page 140)

Quality Counts (page 138)

Many people now have a computer at home, and use it a great deal for a variety of leisure and study activities. There is an opening in the market for an improved method of housing all the different components that make up a computer system.

Design and make a suitable workstation for the home computer user.

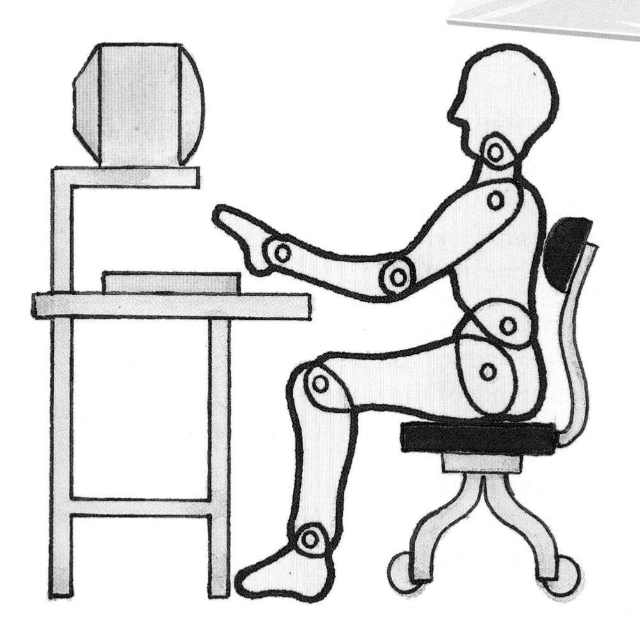

Investigation

Make a detailed study of the components which are necessary for a home computer system.

▷ What are the overall measurements of the items that typically make up a computer system?
▷ How do they need to be arranged?
▷ What extra items may need to be added in the future?

Collect examples of the current special work stations designed for computers. Look particularly at the important ergonomic considerations.

▷ What are the best heights to sit and work at?
▷ How close should the monitor be?
▷ How far is it to reach to switch the system on or off?

Summarise your findings and highlight the key features for your design specification.

First Thoughts

Sketch a number of ideas that you feel are possible ways in which you can solve the problem.

▷ Sketch ergonomically sound shapes that bring different parts of the computer system to within easy reach.
▷ Does the shape of the table surface have to be rectangular? Experiment by changing its shape.
▷ What different arrangements can the various items be placed in on the unit?
▷ Consider the range of possible leg and rail sizes and positions. Which structures will provide the most stable support?
▷ Which combinations of materials can you use?

Decide which are the best aspects of your designs and write down why you want to take these ideas further.

Try to use a range of drawing techniques and add exploded views and assembly drawings as necessary.

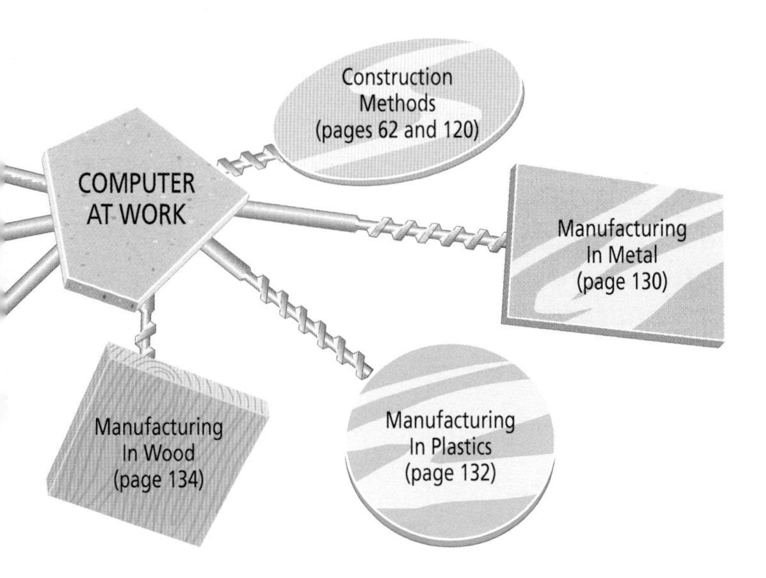

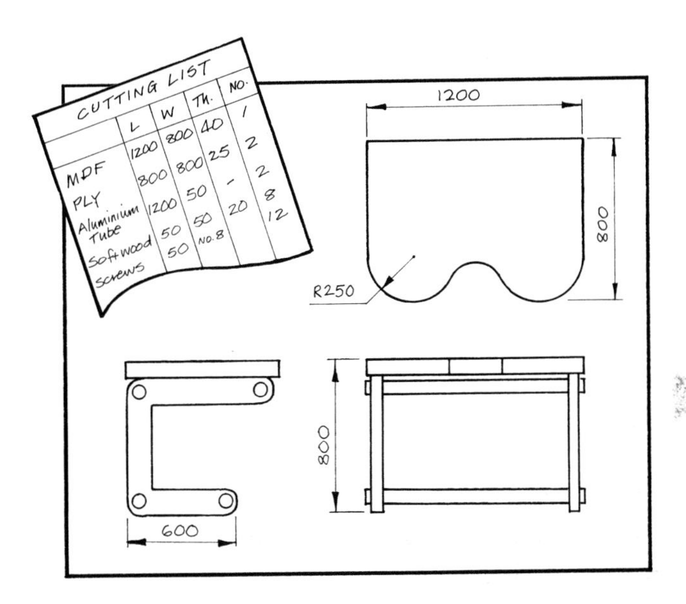

Refining the Design

Begin by focusing on what you have identified as the best aspects of the initial designs and ensure that these are incorporated in the development of your work.

- ▷ Decide on the type of construction that best suits the materials selected.
- ▷ How will various fixtures and fittings be used?
- ▷ What are the best ways of designing the work station if it was to be sold as a self-assembly flat-pack?

Make a scale prototype to see what your design looks like in 3D and to ensure there are no weaknesses in the structure, shape and size.

Remember to prepare:

- ▷ a final orthographic drawing to show overall sizes
- ▷ a list of materials and fixings that you need.

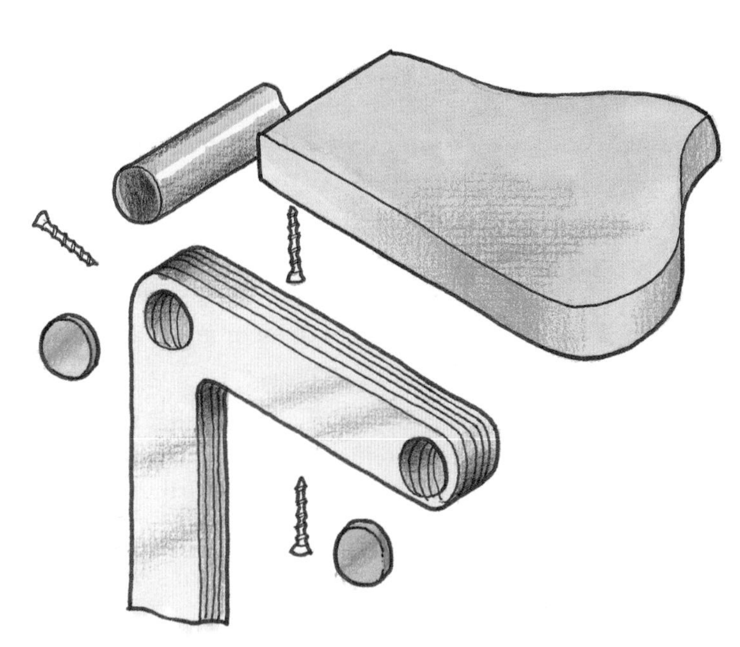

Planning the Making and Testing

Are you sure you have the skills required to make your design? If you have not tried out some of the construction techniques needed then now is the time to practice. Ensure that all hand and power tools are used safely.

You will need to plan the final sequence of making carefully. Include procedures that will ensure everything is made to the highest quality.

Test out your design on a group of people. Take some measurements to check how well it fits them ergonomically.

Designing for Manufacture

- ▷ How might the design, materials and/or production methods need to be changed to make manufacture easier?
- ▷ What jigs or other aids could be used to make the manufacturing process easier?
- ▷ What machine tools could be used to ensure greater accuracy and speed?
- ▷ Which components will need to be checked for quality and safety?

Prepare a short illustrated report to cover these points.

Remember to write your final evaluation.

Learning is Fun

Ergonomics
(page 80)

Combining Colours
and Textures
(page 116)

Fine Finishes
(page 104)

A local play group leader has approached your school to investigate the possibilities of designing and making an educational toy or game. If possible it should interest and motivate children with special educational needs.

SAFETY FIRST!

Remember that safety is an important consideration especially when designing for young children and you will need to respond to existing safety standards.

Investigation

Begin by identifying the types of play young children of this age range engage in. This may be done by visiting a local playgroup or by going to an *Early Learning Centre* shop to see what is available.

You may find it helpful to photograph certain key features, or sketch existing ideas or note important aspects of play activities.

- ▷ What do children seem to enjoy the most?
- ▷ What features seem to make some things more interesting and appealing?
- ▷ What things are children learning about, e.g. colours, shapes, the alphabet, numbers, etc?
- ▷ Are there toys or games that are not played with?
- ▷ Are there things that attract boys more than girls?
- ▷ Are there already any toys and games for those in the group who have special educational needs?

Investigate costs, materials and the function of the activity by analysing existing ideas. These can be collected from a number of different sources.

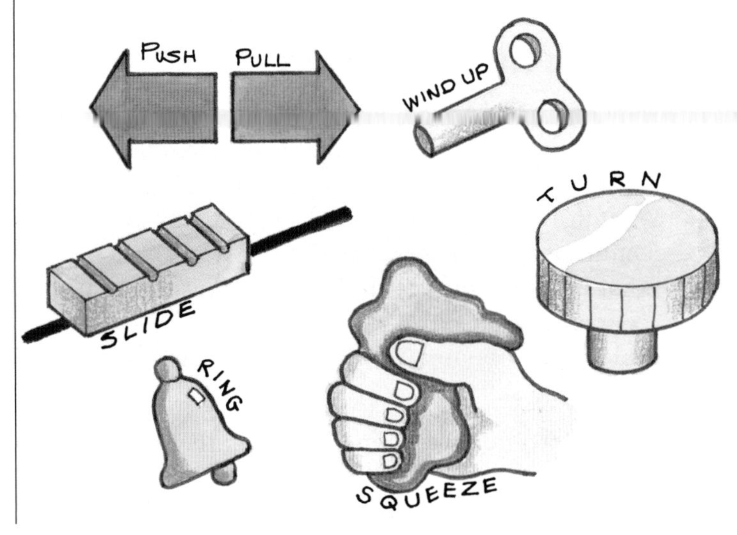

Design Specification

Write a clear specification for the educational toy or game that you are going to design and make. Make sure that you use your research to focus the statements on issues such as safety, materials, cost, needs of the child and how the design will stimulate learning.

First Thoughts

You may want to base your initial designs on some of the existing ideas you gathered in the research. Try to sketch ideas using a range of drawing techniques.

Include a wide range of possible ideas at this stage and develop designs that perhaps move, can be built on, can be joined together, have a feel to them or are of a striking colour.

Add annotation to describe materials and what the design will do. You need to justify your thinking and to give valid reasons for developing an idea further.

It is worth handling some of the materials that you could use in your design and experiment with forms of construction and possible processes.

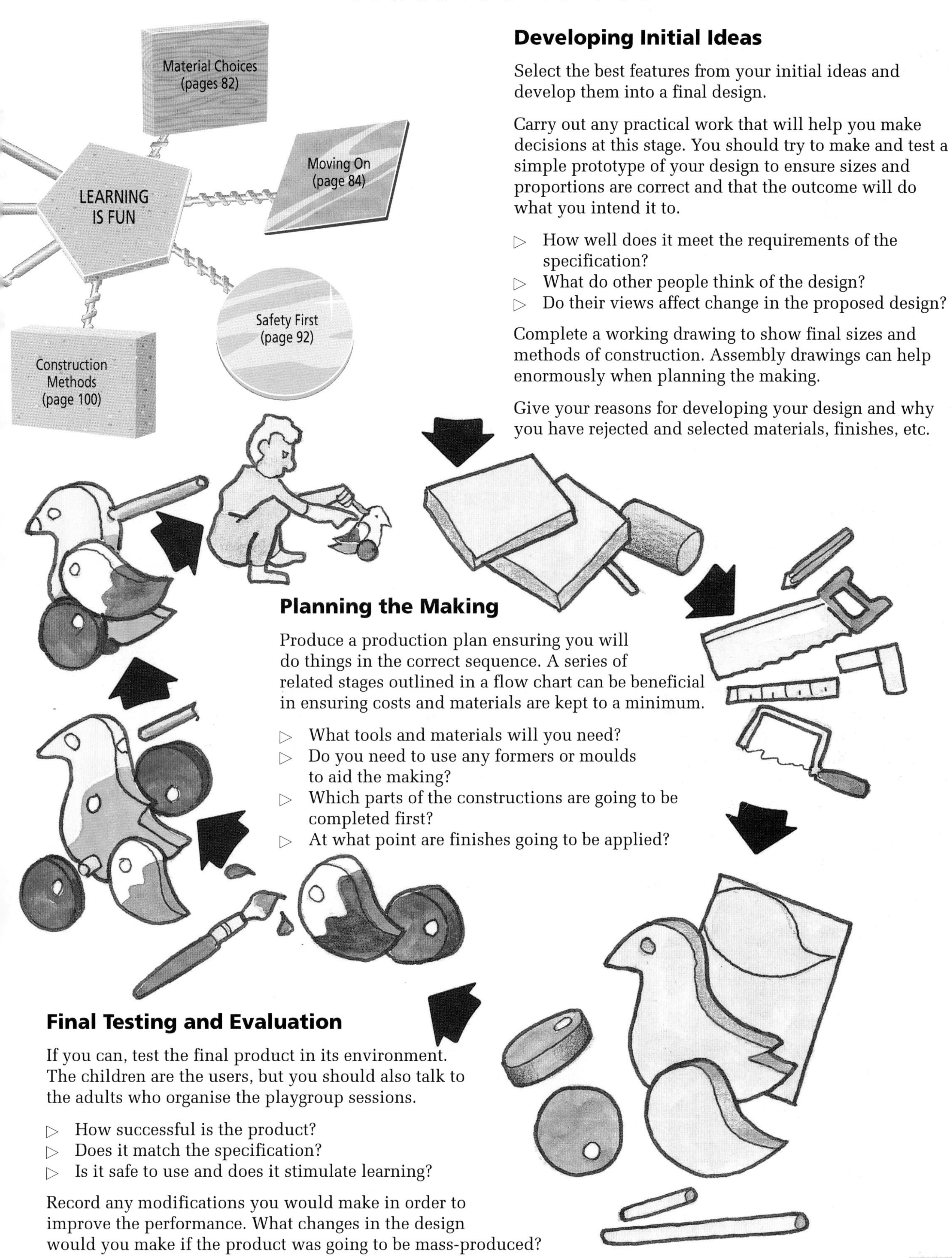

Developing Initial Ideas

Select the best features from your initial ideas and develop them into a final design.

Carry out any practical work that will help you make decisions at this stage. You should try to make and test a simple prototype of your design to ensure sizes and proportions are correct and that the outcome will do what you intend it to.

- ▷ How well does it meet the requirements of the specification?
- ▷ What do other people think of the design?
- ▷ Do their views affect change in the proposed design?

Complete a working drawing to show final sizes and methods of construction. Assembly drawings can help enormously when planning the making.

Give your reasons for developing your design and why you have rejected and selected materials, finishes, etc.

Planning the Making

Produce a production plan ensuring you will do things in the correct sequence. A series of related stages outlined in a flow chart can be beneficial in ensuring costs and materials are kept to a minimum.

- ▷ What tools and materials will you need?
- ▷ Do you need to use any formers or moulds to aid the making?
- ▷ Which parts of the constructions are going to be completed first?
- ▷ At what point are finishes going to be applied?

Final Testing and Evaluation

If you can, test the final product in its environment. The children are the users, but you should also talk to the adults who organise the playgroup sessions.

- ▷ How successful is the product?
- ▷ Does it match the specification?
- ▷ Is it safe to use and does it stimulate learning?

Record any modifications you would make in order to improve the performance. What changes in the design would you make if the product was going to be mass-produced?

New Lamps for Old

A modern light is a great deal more than a simple bulb and a shade. The type of light bulb, the way it is held in position and the colour and shape of the shade all contribute to the design of a lighting device.

Can you design and make a lighting device for a particular purpose or environment? It could be a table lamp, a ceiling lamp, a desk lamp, a wall lamp or a floor lamp.

Manufacturing
(page 130)

Combining
Colours and
Textures
(page 116)

Investigation

Make a study of a wide range of different lighting devices found in the home and the office. In particular look out for:

- different types and sizes of light bulbs (e.g. tungsten and halogen)
- ways of making the device adjustable
- the materials used for the structure and shade
- the methods of construction used.

What can you find out about:

- the levels of lighting needed for particular situations?
- the types, ventilation and running costs of different bulbs?

See if you can discover something about the history of lighting:

- who invented the light bulb and when?
- the lighting designs of Charles Rennie Mackintosh, Louis Comfort Tiffany, Josef Hoffmann and Frank Lloyd Wright
- when and where the anglepoise lamp was designed
- what does the Tizio light look like, and how does it work?

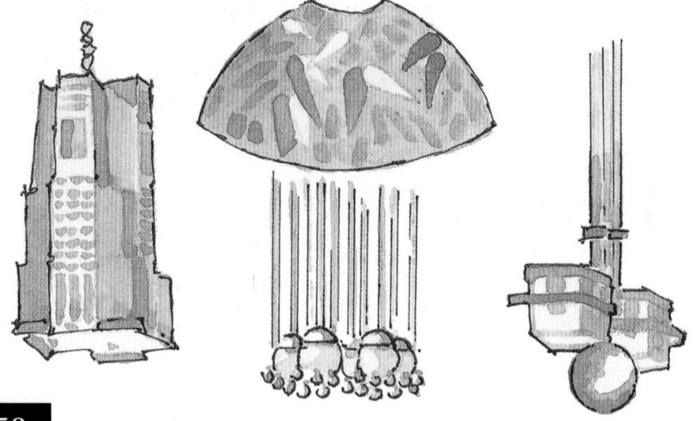

SAFETY FIRST!

To undertake this project you must be familiar with the requirements of electrical safety. Lighting devices which are incorrectly wired up can be extremely dangerous. You may only use mains current with the permission of your teacher. A 12 volt DC supply is much safer.

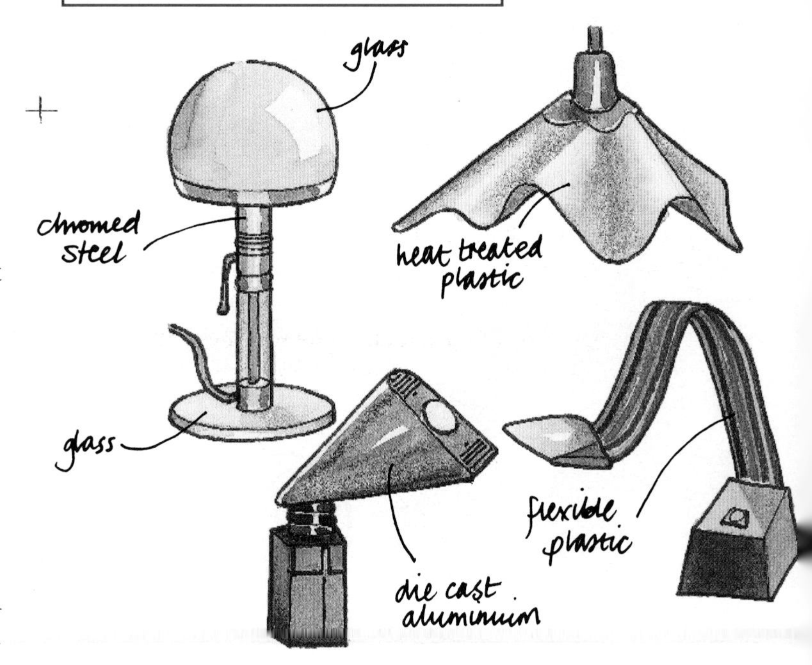

Design Specification

At the end of your investigation you will need to prepare a specification which covers details of things like performance, safety, appearance, maintenance and cost. Remember to include numerical data where possible.

You should also state clearly what particular purpose or environment your light is to be designed for. Will it be for reading, writing, eating, cooking, etc?

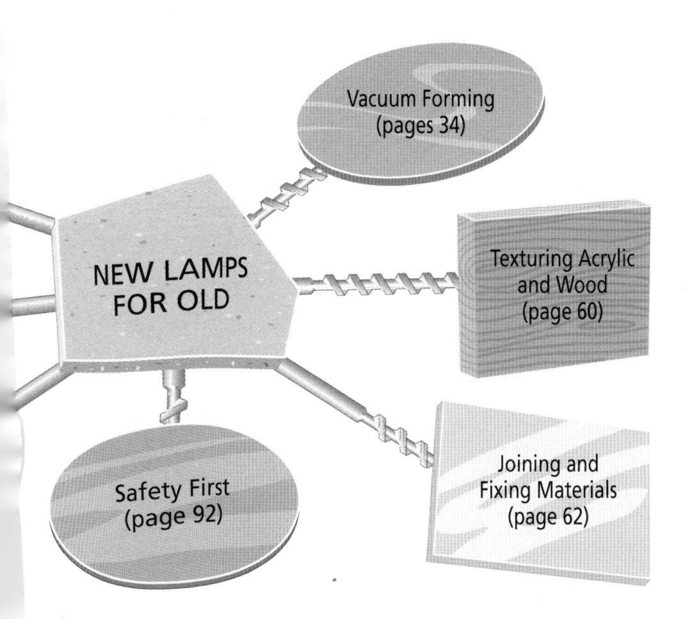

First Thoughts

As you explore some initial ideas, you should consider the following:

▷ How can adjustment be provided?
▷ How can glare from the bulb be reduced?
▷ How will the unit be prevented from falling over?
▷ What switching devices are available, and where could they be placed?
▷ Will it be easy to change the bulb when it needs replacing?
▷ What unusual materials such as ceramics, coloured glass, or textiles could be used?
▷ Which colours, textures and patterns will be most effective?
▷ How might the electrical cable be hidden or be incorporated into the design?

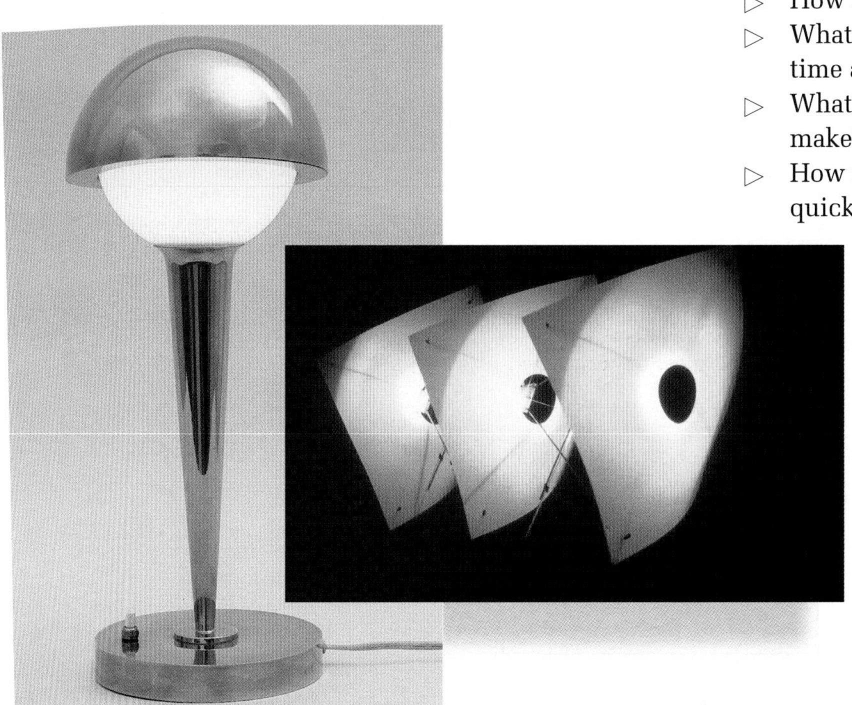

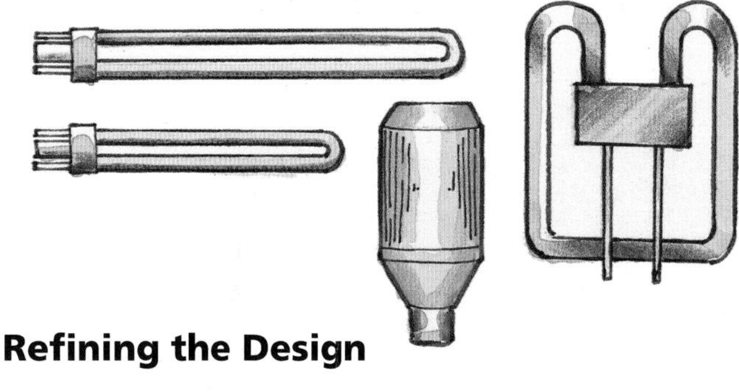

Refining the Design

As you get down to considering details you will need to construct some prototypes to help you test and evaluate your ideas in 3D. You might want to:

▷ build something full-size to check out structural and constructional details and to experiment with different shades
▷ make a part of the design to test out moving or adjustable components
▷ make a scale model to help evaluate the proportions, visual elements and any decorative details.

Planning and Making It!

Your final design will need to be presented as a series of working drawings showing all the details of how it will be made.

You will also need to produce a time plan, though you may wish to modify this throughout the making process.

Planning for Manufacture

▷ How could you make a batch of 1000 units?
▷ How much would the materials cost?
▷ What production methods would you use to save time and money?
▷ What quality control checks would you need to make?
▷ How might the design need to be changed to make it quicker, easier and cheaper to manufacture?

Final Testing and Evaluation

Devise a series of tests to check how well your design performs against the original specification.

Refer to the results of these tests when you write your evaluation. As well as discussing the good and bad points about your design, you should also refer to the process you used to develop and finalise your ideas.

Index